AS-Level Biology AQA A

The Revision Guide

Editors:
Becky May, Kate Redmond.

Contributors:
Gloria Barnett, Claire Charlton, Martin Chester, Barbara Green, Anna-Fe Guy, Dominic Hall, Gemma Hallam, Stephen Phillips, Claire Reed, Katherine Reed, Adrian Schmit, Emma Singleton, Sharon Watson.

Proofreaders:
Ben Aldiss, Vanessa Aris, James Foster, Tom Trust.

Published by Coordination Group Publications Ltd.

ISBN-10: 1 84146 955 6
ISBN-13: 978 1 84146 955 3

Groovy website: www.cgpbooks.co.uk
Jolly bits of clipart from CorelDRAW®
Printed by Elanders Hindson Ltd, Newcastle upon Tyne.

Contents

Section One — Cells

Cells and Microscopy 2
Functions of Organelles 4
Cell Fractionation 5
Plasma Membranes 6
Transport Across Membranes 7

Section Two — Molecules

Carbohydrates 10
Proteins 12
Lipids 14
Biochemical Tests for Molecules 16
Chromatography 17
Action of Enzymes 18
Factors that Affect Enzyme Activity 20

Section Three — Systems

Tissues 22
Surface Area to Volume Ratio 23
Organs and Blood Transport 24
Tissue Fluid 25
Lungs and Ventilation 26
The Heart and the Cardiac Cycle 28
Effects of Exercise 30

Section Four— Making Use of Biology

Isolation of Enzymes 32
Mitosis and the Cell Cycle 34
Meiosis 36
Basic Structure of DNA and RNA 38
Replication of DNA 39
The Genetic Code 40
Types of RNA 41
Protein Synthesis 42
Recombinant DNA 44
The Ethics of Genetic Engineering 47
Immunology 48
Blood Groups 49
Genetic Fingerprinting 50
Polymerase Chain Reaction 51
Adaptations of Cereals 52
Controlling the Abiotic Environment 53
Fertilisers 54
Pesticides 55
Reproduction and its Hormonal Control 56
Manipulation and Control of Reproduction 57

Answers 58

Index 64

Cells and Microscopy

Woohoo — cells. Not the most original way to start a biology book, but hey. We're all made of cells, so you can't knock 'em really. Plus, you need to learn about microscopes, otherwise cells will forever remain meaningless colourful splodges.

There are Two Types of Cell — Prokaryotic and Eukaryotic

Prokaryotic cells are **more simple** than eukaryotic cells. Prokaryotes include **bacteria** and **blue-green algae**. **Eukaryotic** cells are more complex, and include all **animal and plant cells**.

PROKARYOTES	EUKARYOTES
Extremely small cells (0.5-3.0 μm diameter)	Larger cells (20-40 μm diameter)
DNA is circular	DNA is linear
No nucleus — DNA free in cytoplasm	Nucleus present — DNA is inside nucleus
Cell wall made of a polysaccharide, but not cellulose or chitin	No cell wall (in animals), cellulose cell wall (in plants) or chitin cell wall (in fungi)
Few organelles, no mitochondria	Many organelles, mitochondria present
Small ribosomes	Larger ribosomes
Example: *E. coli* bacterium	Example: Human liver cell

(See the section below for pictures of **eukaryotic plant and animal cells**.)

A Typical Prokaryotic Cell

plasma membrane
DNA
cell wall
flagellum (helps bacterium move)
ribosome
plasmid
capsule

The capsule is a gelatinous (slimy) layer found on the outside of prokaryotic cells. It protects cells and stops them drying out.

Light Microscopes Show Cell Structure

If you just want to see the **general structure of a cell**, then light microscopes are fine. But even with the best light microscopes, you can't see most of the organelles in the cell. You can see the larger organelles, like the **nucleus**, but none of the internal details.

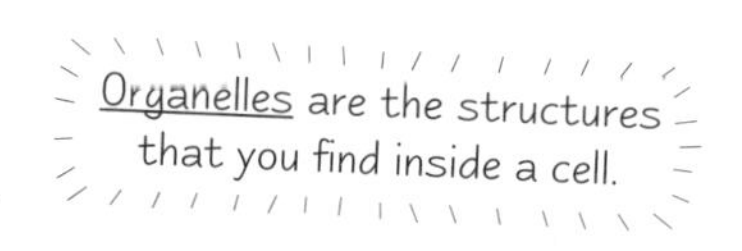

Liver cells seen under a light microscope:

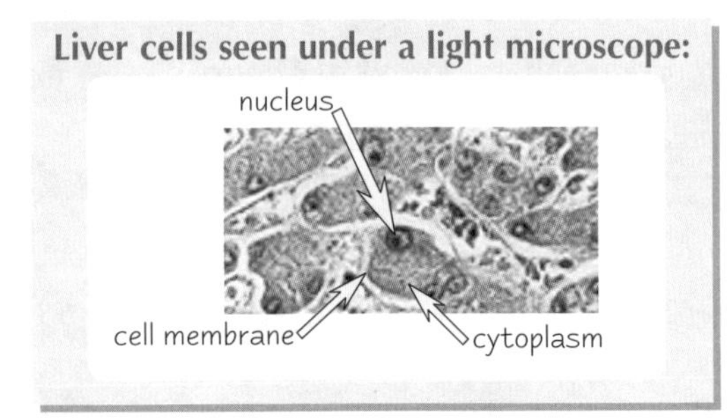

General animal and plant cell, as if seen under a light microscope:

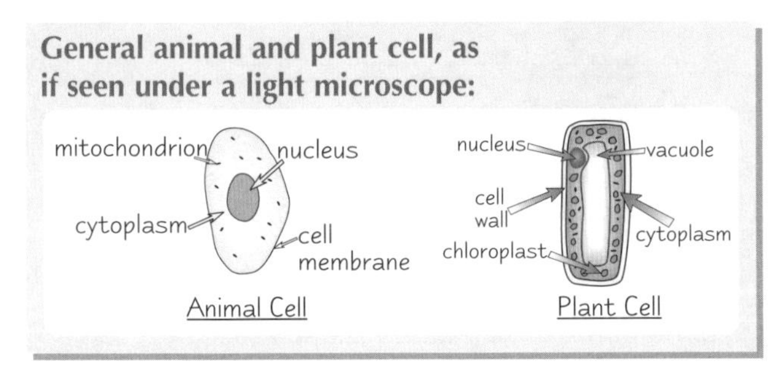

Under a light microscope, you can just about make out that the membrane along the top surface of some animal cells is sometimes **folded** to form **microvilli** (also called a '**brush border**'). This **increases surface area** for **absorption** of substances. A good example is the **epithelial cells** that line the **small intestine**. The microvilli help them to efficiently absorb the products of digestion. NB, microvilli are much clearer under an electron microscope.

Epithelial Cell in the Small Intestine

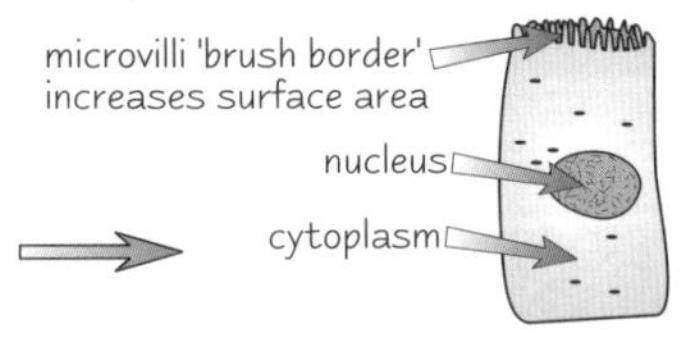

Electron Microscopes Show Cell Ultrastructure

There's not much in a cell that an electron microscope can't see. You can see a cell's **ultrastructure**, which is its **organelles** and the **internal structure** of most of them. Most of what's known about cell structure has been discovered by electron microscope studies. The diagram to the right shows what you can see in an animal cell under an electron microscope. Very pretty indeed.

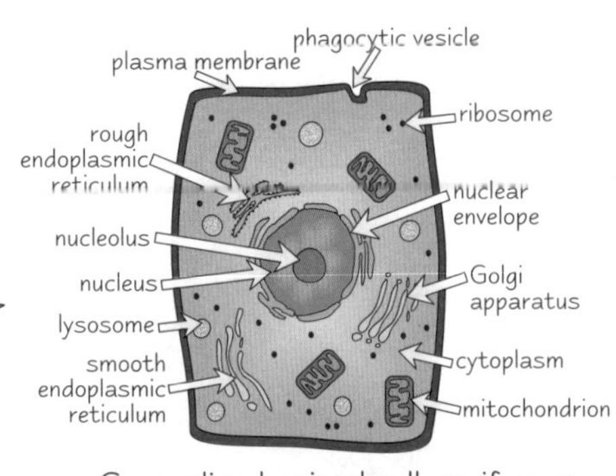

Generalised animal cell, as if seen under an electron microscope

Cells and Microscopy

Magnification is Size, Resolution is Detail

1) **Magnification** is how much bigger the image is than the specimen. It's calculated as: $\frac{\text{length of drawing or photograph}}{\text{length of specimen object}}$
2) **Resolution** is how detailed the image is. More specifically, it's how well a microscope distinguishes between two points that are close together. If a microscope lens can't separate two objects, then increasing the magnification won't help.

electron gun
condenser
air lock/specimen port
specimen position
objective
projector
viewing port
vacuum pump
fluorescent screen
camera chamber

Electron Microscope

- **Light microscopes** have a **lower resolution** than electron microscopes. Decreasing the wavelength of the light increases resolution, but even then a light microscope can only distinguish points 0.2 micrometres (μm) apart.

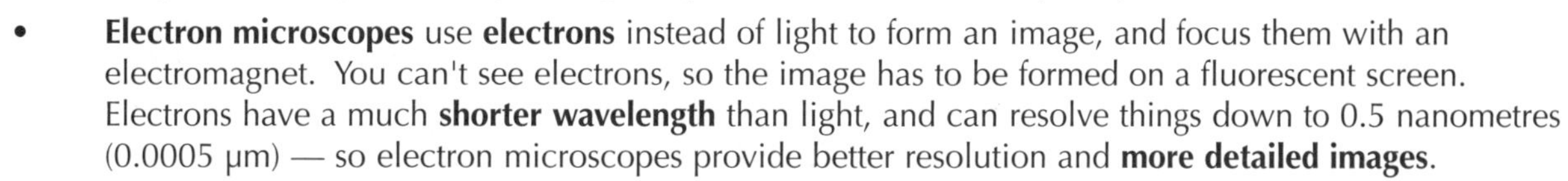

- **Electron microscopes** use **electrons** instead of light to form an image, and focus them with an electromagnet. You can't see electrons, so the image has to be formed on a fluorescent screen. Electrons have a much **shorter wavelength** than light, and can resolve things down to 0.5 nanometres (0.0005 μm) — so electron microscopes provide better resolution and **more detailed images**.

There are two types of electron microscope:

1) **Transmission electron microscope (TEM)** — This uses **electromagnets** to focus a **beam of electrons**, which is then transmitted through the specimen. **Denser** parts of the specimen absorb more electrons, which makes them look **darker** on the image you end up with. TEMs are good because they provide **high resolution images**, but they can only be used on **thin specimens**.
2) **Scanning electron microscope (SEM)** — This scans a beam of electrons across the specimen, and reflected electrons are gathered in a **cathode ray tube**, which forms a **TV image**. The images you end up with show the **surface** of the specimen and they can be **3-D**. SEMs provide **lower resolution images** than TEMs.

Electron Microscopes Have Advantages and Disadvantages

ADVANTAGES OF AN ELECTRON MICROSCOPE	DISADVANTAGES OF AN ELECTRON MICROSCOPE
Gives higher resolution images than a light microscope. Resolution is down to 0.5 nm.	Processing kills living cells
Maximum magnification is higher than it is with light microscopes.	Natural colours can't be seen, whereas colours can be seen with a light microscope.
	Electron microscopes can't be moved around, whereas light microscopes are mobile.
	Very expensive.

A micrometre (μm) is a thousandth of a millimetre. A nanometre (nm) is a thousandth of a micrometre. That's tiny.

Despite the disadvantages of electron microscopes, their advantages are big advantages.

Practice Questions

Q1 Name three differences between eukaryotic and prokaryotic cells.

Q2 Why do electron microscopes have a better resolution than light microscopes?

Q3 What is used to focus an electron microscope?

Q4 Name the two types of electron microscope and describe how each one works.

Exam Question

Q1 Explain the advantages and disadvantages of using an electron microscope rather than a light microscope to study cells. [6 marks]

Cells — tiny little blobs with important jobs to do...

OK, so you've read the first topic and now it's the moment you've been waiting for — yep, it's time to get learning those facts. You need to know the differences between eukaryotic and prokaryotic cells, and between magnification and resolution and between the two types of electron microscopes. And brush your hair — you look like a mess.

Functions of Organelles

Organelles are all the tiny bits and bobs inside a cell that you can only see in detail with an electron microscope. It's cool to think that all these weird and wonderful things live inside our tiny cells.

Cells Contain Organelles

An organelle is a structure found inside a cell — each one has a specific function. Most organelles are surrounded by membranes, which sometimes causes confusion — don't make the mistake of thinking that a diagram of an organelle is a diagram of a whole cell. They're not cells — they're **parts of** cells, see.

ORGANELLE	DIAGRAM	DESCRIPTION	FUNCTION
Cell wall	plasma membrane, cell wall, cytoplasm	A rigid structure that surrounds **plant cells**. It's made mainly of the carbohydrate **cellulose**.	**Supports** plant cells.
Plasma membrane	plasma membrane, cytoplasm	The membrane found on the surface of **animal cells** and just inside the cell wall of **plant cells**. It's made mainly of **protein** and **lipids**.	**Regulates the movement** of substances into and out of the cell. It also has **receptor molecules** on it, so it can respond to chemicals like hormones.
Nucleus	nuclear membrane, nucleolus, nuclear pore, chromatin	A large organelle surrounded by a **nuclear membrane**, which contains many **pores**. The nucleus contains **chromatin** and often a structure called the **nucleolus**.	The **chromatin** contains genetic material (DNA) which **controls cell activities**. The pores allow things like RNA to move between the nucleus and the cytoplasm. The **nucleolus** makes **RNA**.
Lysosome		A **round organelle** surrounded by a **membrane**, with no clear internal structure.	Contains **digestive enzymes**. These can be used to **digest invading cells** or to **destroy the cell** when it needs to be replaced.
Ribosome	small subunit, large subunit	A **very small organelle** either floating free in the cytoplasm or attached to the rough endoplasmic reticulum.	The **site** where **proteins** are made.
Endoplasmic Reticulum	a) b) ribosome, fluid	There are 2 types of endoplasmic reticulum: the **Smooth Endoplasmic Reticulum** (diagram **a**) is a system of membranes enclosing a fluid-filled space; the **Rough Endoplasmic Reticulum** (diagram **b**) is similar, but is **covered in ribosomes**.	The **Smooth Endoplasmic Reticulum** transports **lipids** around the cell. The **Rough Endoplasmic Reticulum** transports **proteins** which have been made in the ribosomes.
Golgi Apparatus	vesicle	A group of smooth endoplasmic reticulum consisting of a series of **flattened sacs**. Vesicles are often seen at the edges of the sacs.	It **packages** substances that are produced by the cell, mainly proteins and glycoproteins. It also **makes lysosomes**.
Vesicle	pinocytic vesicle, cell's plasma membrane, vesicle	A small **fluid-filled** sac in the cytoplasm, surrounded by a membrane.	**Transports substances** to and from the cell via the plasma membrane. Some are formed by the Golgi apparatus, while others (**pinocytic** or **phagocytic** vesicles) are formed at the cell surface.
Mitochondrion	outer membrane, inner membrane, crista, matrix	They are usually oval. They have a **double membrane** — the inner one is folded to form structures called **cristae**. Inside is the **matrix**, which contains enzymes involved in respiration (but, sadly, no Keanu).	The **site of respiration**, where **ATP** is produced. They are found in large numbers in cells that are very active and require a lot of energy.
Chloroplast	stroma, two membranes, granum (plural = grana), lamella (plural = lamellae)	A small, **flattened** structure found in **plant cells**. It's surrounded by a **double membrane**, and also has **thylakoid membranes** inside. These are stacked up in some parts to form **grana**. Grana are linked together by **lamellae**.	The **site** where **photosynthesis** takes place. The light-dependent reaction of photosynthesis happens in the **grana**, and the light-independent reaction of photosynthesis happens in the **stroma**.

Cell Fractionation

Cell Fractionation Sorts Organelles

To separate a type of organelle from all the others in a cell, you use **cell-fractionation** to **break-up** the cell. Then you **spin** the broken-up cell at high speed (this is called **ultracentrifugation**) to separate out the different organelles.

Differential centrifugation (centrifuging at different speeds) is used to isolate each organelle so it can be extracted separately:

Isotonic means equal concentration. The concentration is equal to the concentration of the fluids in the organelles.

1) First, the cells are '**homogenised**' (mashed up) in **ice-cold isotonic buffer solution**. The **low temperature** prevents protein-digesting enzymes **digesting** the organelles. The **buffer** keeps the **pH constant** and the **isotonic solution** stops the organelles taking in lots of water via **osmosis** and bursting.
2) The cell fragments are poured into a **tube**. The tube is put into a **centrifuge** (with the **bottom** of the tube facing **outwards**, so heavy stuff gets flung outwards and ends up at the bottom of the tube), and is spun at a **low speed**. **Cell debris**, like the cells walls in plant cells, gets flung to the bottom of the tube by the centrifuge. It forms a thick sediment at the bottom, which is called a **pellet**. The rest of the organelles stay suspended in the **supernatant** (the fluid above the sediment).
3) The supernatant is **drained off**, poured into **another tube**, and spun in the centrifuge at a **higher speed**. The heavier organelles like the nuclei form a pellet at the bottom of the tube. The supernatant containing the rest of the organelles is drained off and spun in the centrifuge at an even higher speed.
4) This process is repeated at higher and higher speeds, until all the organelles are **separated out**. Each time, the pellets at the bottom of the tube are made up of lighter and lighter organelles.

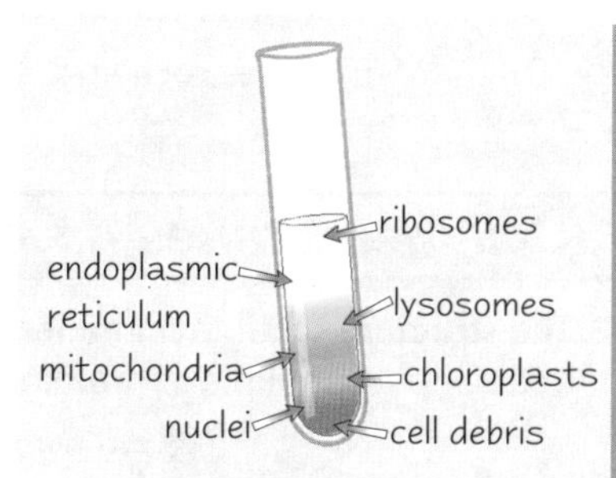

The organelles are separated in order of mass (from heaviest to lightest) — this order is usually: cell debris, then nuclei, then chloroplasts, then mitochondria, then lysosomes, then endoplasmic reticulum, and finally ribosomes.

Practice Questions

Q1 Name two organelles found only in plant cells.

Q2 What is the function of lysosomes?

Q3 Explain the differences between rough and smooth endoplasmic reticulum.

Q4 State the name of the process that's used to separate organelles.

Q5 Explain what the term "isotonic" means.

Exam Questions

Q1 The presence and number of specific organelles can give an indication of a cell's function. Give THREE examples of this, naming the organelles concerned and stating their function. [9 marks]

Q2 a) Identify these two organelles seen in an electron micrograph, from the descriptions given below.

(i) A sausage-shaped organelle surrounded by a double membrane. The inner membrane is folded and projects into the inner space, which is filled with a grainy material.

(ii) A collection of flattened membrane 'bags' arranged roughly parallel to one another. Small circular structures are seen at the edges of these 'bags'. [2 marks]

b) State the function of the two organelles that you have identified. [2 marks]

Q3 Describe how cells are prepared for differential centrifugation. [5 marks]

Organs and organelles — 'his and her' biology terms...

You need to know the names and functions of all the organelles and also what they look like under the microscope. Differential centrifugation might have possibly one of the most poncy names in biology (and that's saying something) — but you've still got to learn all about it, otherwise you can say bye bye to easy marks.

Plasma Membranes

This page is all about the structure of cell membranes. Try and contain your excitement when you read about the fluid mosaic model — there have been some nasty cases of extreme over-excitement in the past.

Membranes Control What **Passes Through** Them

Cells and many of the **organelles** inside them are surrounded by **membranes**. Membranes have a **range of functions**.

1) **Membranes around organelles** divide the cell up into **different compartments** to make the different **functions more efficient** — e.g. the substances needed for **respiration** (like enzymes) are kept together inside **mitochondria**.
2) Membranes control **which substances enter and leave** a cell or organelle.
3) Membranes **recognise** specific chemical substances and other cells.

Cell Membranes have a **'Fluid Mosaic' Structure**

The **structure** of all **membranes** is basically the same. They are composed of **lipids** (mainly phospholipids), **proteins** and **carbohydrates** (usually attached to proteins or lipids).

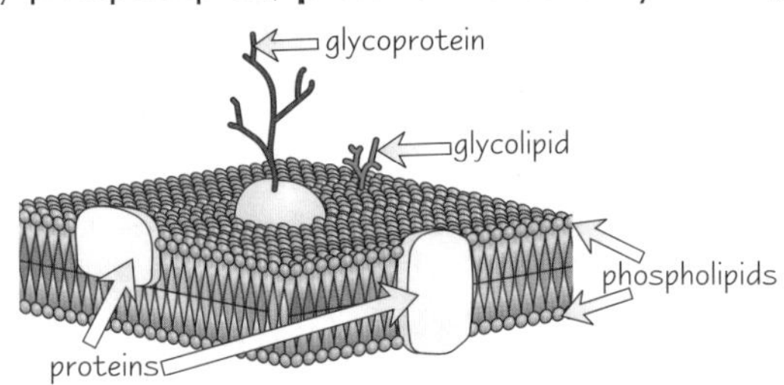

In 1972, the **fluid mosaic model** was suggested to describe the arrangement of molecules in the membrane. In the model, **phospholipid molecules** form a continuous, double layer (**bilayer**). This layer is 'fluid' because the phospholipids are constantly moving. **Protein molecules** are scattered through the layer, like tiles in a **mosaic**.

Phospholipids Can Form **Bilayers**

Phospholipids consist of a **glycerol molecule** plus **two molecules** of **fatty acid** and a **phosphate group** (see p.15).

1) The **phosphate / glycerol head** is **hydrophilic** — it attracts water. The **fatty acid tails** are **hydrophobic** — they repel water.
2) In **aqueous** (**water-based**) **solutions** phospholipids automatically arrange themselves into a **double layer** so that the **hydrophobic tails** pack together **inside the layer** away from the water, and the **hydrophilic heads face outwards** into the aqueous solutions.

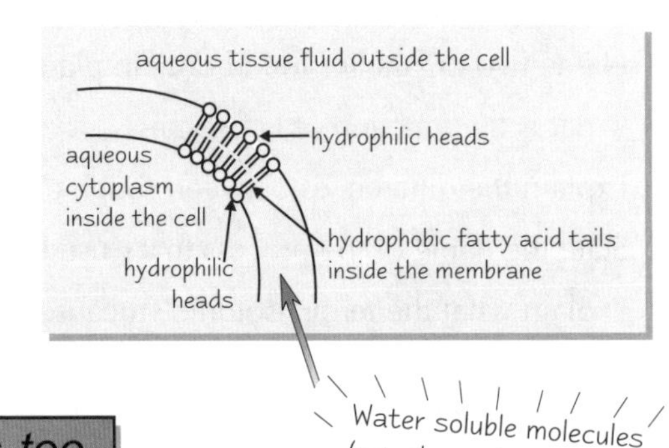

Water soluble molecules (e.g. glucose) can't pass through the fatty, hydrophobic interior of the membrane.

Membranes Contain loads of **Other Molecules** too

1) **Cholesterol** molecules are often found in between phospholipid molecules — they make the membrane **less fluid** and **more stable**.
2) **Glycoproteins** are proteins with **carbohydrates attached**. They're found on the **cell membrane surface** and are important for **cell recognition** — e.g. some act as **antigens**, meaning they're recognised by white blood cells (see p.48 for more on this).
3) **Channel proteins** form a tiny **gap** in the membrane to allow water-soluble molecules and ions through by **diffusion**.
4) **Carrier proteins** carry water-soluble molecules and ions through the membrane by **active transport** and **facilitated diffusion** (see p.8).
5) **Receptor proteins** are found on the membrane surface — they recognise and bind to **specific molecules** (e.g. hormones).

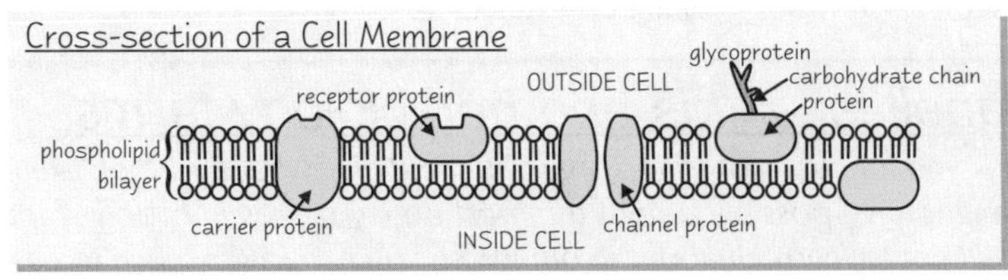

Transport Across Membranes

*There are six methods of transport across a cell membrane. You need to learn all six — **diffusion, osmosis, facilitated diffusion, active transport, endocytosis** and **exocytosis**. It's a big topic, so the next three pages are dedicated to it.*

1) Diffusion is the Passive Movement of Particles

1) Diffusion is the net movement of particles (molecules or ions) from an area of **higher concentration** to an area of **lower concentration**. This continues until particles are **evenly distributed** throughout the liquid or gas.
2) Diffusion is described as a **passive process** because **no energy** is needed for it to happen.
3) Diffusion can happen **across cell membranes**, as long as particles can **move freely** through the membrane. E.g. water, oxygen and carbon dioxide molecules are small enough to pass easily through pores in the membrane.

The Speed of Diffusion Depends on Several Factors

1) The **concentration gradient** is the difference in concentration between an area of **higher concentration** and an area of **lower concentration**. Particles diffuse **faster** when there is a **big concentration gradient**.
2) The **shorter** the **distance** the particles have to travel, the **faster** the rate of diffusion.
3) The **larger** the **surface area** of the cell membrane, the **faster** the rate of diffusion.

The rate at which a substance diffuses can be worked out using **Fick's law**:

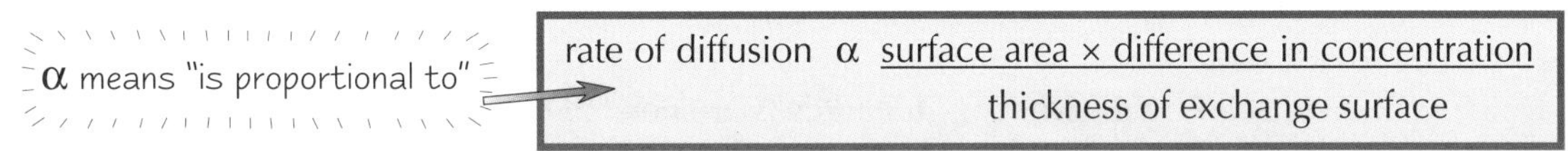

$$\text{rate of diffusion} \;\alpha\; \frac{\text{surface area} \times \text{difference in concentration}}{\text{thickness of exchange surface}}$$

Water Potential is the Ability of Water Molecules to Move

1) **Water potential** is the potential (likelihood) of water molecules to diffuse out of a solution.
2) Water potential is represented by the symbol ψ. It's measured in **kilopascals** (kPa).
3) The water potential of **pure water** is **zero kilopascals**. All solutions have a **lower** water potential than pure water, so their water potentials are always **negative**.
4) Water molecules **diffuse** from solutions with a **higher water potential** (**hypotonic** solutions) to solutions with a **lower water potential** (**hypertonic** solutions) until both solutions have the same water potential (they are **isotonic**).
5) Water potential is **important for osmosis** — see over the page for more on this. (Oooh, the anticip......ation.)

Practice Questions

Q1 Give three functions of cell membranes.

Q2 Which types of molecules are carbohydrate molecules usually attached to in the cell membrane?

Q3 Which part of a phospholipid molecule is hydrophobic?

Q4 What is meant by the term "concentration gradient"?

Exam Questions

Q1 Describe the role of phospholipids in controlling the passage of water soluble molecules through the cell membrane. [2 marks]

Q2 The diagram shows three cells, each with a different water potential.
Draw arrows onto the diagram, showing the flow of water between the three cells. [3 marks]

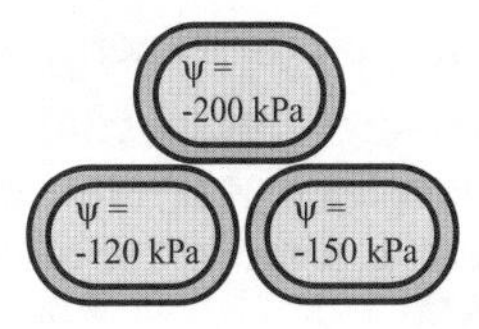

Membranes actually **are** all around...

The cell membrane is a complex structure — but then it has to be, 'cos it's the line of defence between a cell's contents and all the big bad molecules outside. Don't confuse the cell membrane with the cell wall (found in plant cells). The cell membrane controls what substances enter and leave the cell whereas the cell wall provides structural support.

Transport Across Membranes

2) Osmosis is a Particular Kind of Diffusion Involving Water Molecules

1) Osmosis is when **water molecules** diffuse through a **partially permeable membrane** from an area of **higher water potential** (i.e. higher concentration of water molecules) to an area of **lower water potential**.
2) A **partially permeable membrane** allows some molecules through it, but not all. Water molecules are small and can diffuse through easily but large solute molecules can't.
3) Water molecules will diffuse **both ways** through the membrane — but the **net movement** will be to the side with a **lower concentration of water molecules**.

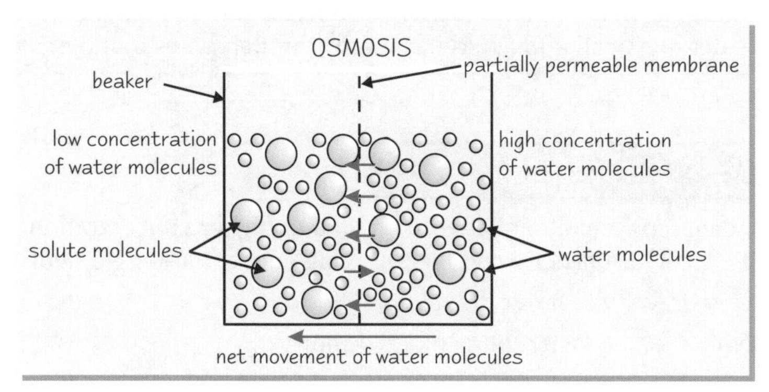

Partially permeable membranes can be useful at sea.

Water moves into and out of **plant cells** by osmosis. **Sugars and ions** inside a plant cell lower the water potential inside the cell, so water moves **into** the cell by osmosis. This water makes the plant cell expand and push against the cell wall, which makes the cell **turgid** (swollen). Turgid cells are needed to keep the plant **upright** and held **firmly in the ground**. The opposite of turgid is **flaccid**, where cells go floppy because water moves out of them by osmosis. Flaccid cells are bad news for a plant — it's what's happening when a plant **wilts**.

3) Facilitated Diffusion uses Carrier Proteins and Channel Proteins

Some **larger molecules** (e.g. amino acids, glucose) and **charged atoms** (e.g. sodium ions) can't diffuse through the phospholipid bilayer of the cell membrane themselves. Instead they diffuse through **carrier proteins** or **channel proteins** in the cell membrane. This is called **facilitated diffusion**.

1) Channel proteins form **pores** through the membrane for charged particles to diffuse through.
2) Carrier proteins **change shape** to move large molecules into and out of the cell:

The carrier proteins in the cell membrane have **specific shapes** — so specific carrier proteins can only facilitate the diffusion of specific molecules. Facilitated diffusion can only move particles along a **concentration gradient**, from a higher to a lower concentration. It **doesn't** use any **energy**.

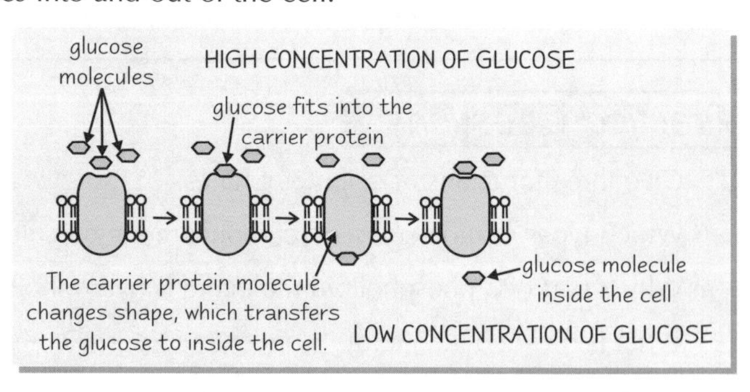

4) Active Transport Moves Substances Against a Concentration Gradient

1) Active transport uses **energy** to move **molecules** and **ions** across cell membranes, **against** a **concentration gradient**.
2) Molecules attach to **specific carrier proteins** (sometimes called 'pumps') in the **cell membrane**, then **molecules of ATP** (adenosine triphosphate) provide the energy to change the shape of the protein and move the molecules across the membrane.

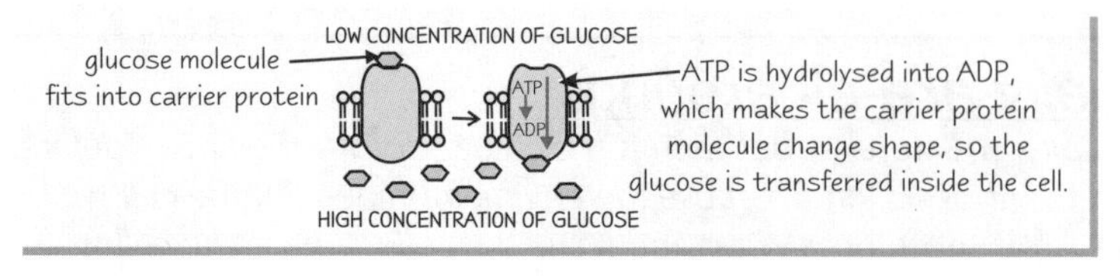

Transport Across Membranes

5 Materials can be Taken into Cells by Endocytosis

Endocytosis is when a cell takes in substances by surrounding them with a section of the cell membrane to form a small vacuole called a **vesicle**.

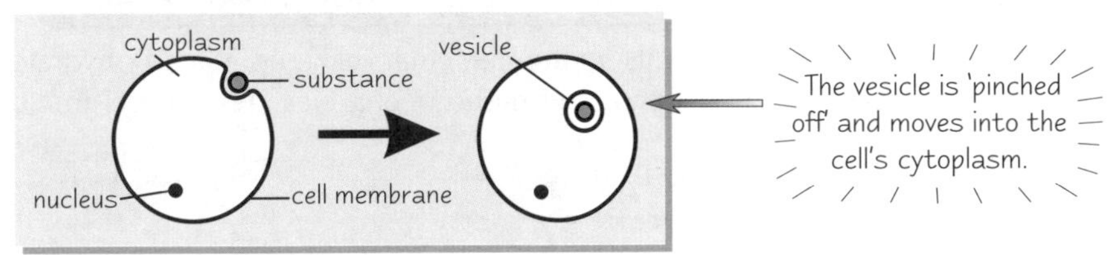

There are 2 types of endocytosis:

1) **Phagocytosis** is when solid particles or whole cells are brought into the cell. The contents of the vesicle are **digested** by **enzymes** secreted from **lysosomes**. Molecules and ions then **diffuse** out of the vesicle into the cell's **cytoplasm**.
2) **Pinocytosis** is similar to phagocytosis — but **liquid** is taken into the cell.

6 Materials can be Removed from Cells by Exocytosis

Materials are **secreted out** of cells by **exocytosis.**

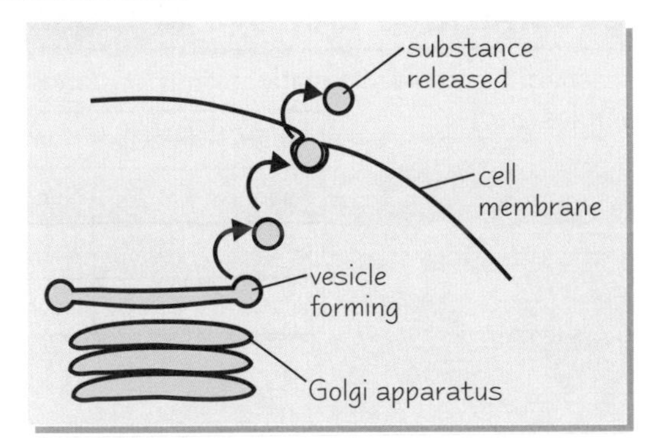

1) **Substances produced by the cell** move through the **endoplasmic reticulum** to the **Golgi body**.
2) **Vesicles** pinch off from the sacs of the Golgi body and move towards the cell membrane. They **merge** with the **cell membrane** and **release** their contents outside of the cell.
3) Digestive enzymes, hormones, mucus and milk are secreted by **exocytosis**.

Practice Questions

Q1 What is a partially permeable membrane?

Q2 Why do plants need turgid cells?

Q3 What kind of proteins are involved in facilitated diffusion?

Q4 Which type of transport through a membrane needs energy?

Q5 What molecule supplies this energy?

Q6 Give two examples of substances secreted by exocytosis.

Exam Questions

Q1 a) Give a definition of osmosis. [2 marks]

b) Explain the significance of osmosis in keeping plants upright. [3 marks]

Q2 a) In terms of water potential, explain how water moves from the soil into a root hair cell. [3 marks]

b) Explain the process that allows the active uptake of ions into root hair cells. [4 marks]

A little less conversation, a little more exocytosis, baby...

Phew, the end of a mammoth topic on transport through the cell membrane — so now you can move on and forget it ever happened. Just kidding (I should be doing stand-up, no really) — now you need to go back over it and check you know the details. Learn the differences between similar terms, like phagocytosis and pinocytosis.

Carbohydrates

All carbohydrates contain only carbon, hydrogen and oxygen. Carbohydrates are dead important chemicals — for a start they're the main energy supply in living organisms and some of them, like cellulose, have an important structural role.

Carbohydrates are Made from **Monosaccharides**

Sugar molecules are the basic units (**monomers**) that make up all carbohydrates. A single sugar molecule is called a **monosaccharide**. Examples of monosaccharides are glucose, fructose and galactose.

There are **two types of glucose** — **alpha** (α) and **beta** (β) glucose. You need to know how their molecules are arranged slightly differently. This has important effects on their **properties** and **functions** (see p.11).

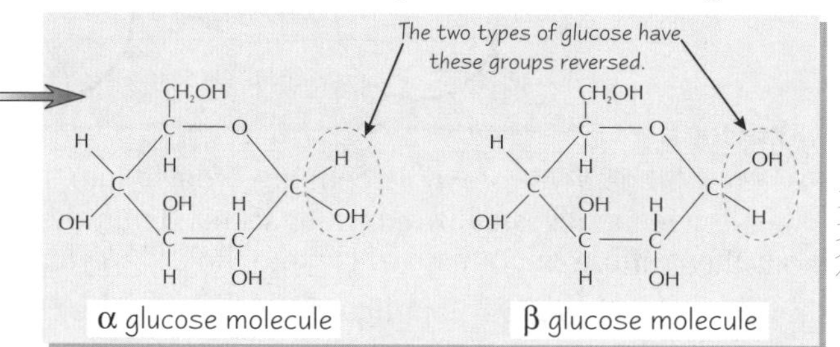

Remember, beta glucose has the H on the bottom as you look at the structural diagram.

Disaccharides are **Two Monosaccharides** Joined Together

Disaccharides are sugars made from two monosaccharide sugar molecules stuck together. Examples are:

DISACCHARIDE	MONOSACCHARIDES IT'S MADE UP OF
maltose	glucose + glucose
sucrose	glucose + fructose
lactose	glucose + galactose

Extensive scientific research revealed an irreversible bond joining sugars to Pollyanna's gob.

Glycosidic Bonds Join Sugars Together

Sugars are held together by **glycosidic bonds**. When the sugars join, a molecule of water is squeezed out. This is called a **condensation reaction**.

If you're asked to show a condensation reaction in an exam, don't forget to put the water molecule in as a product.

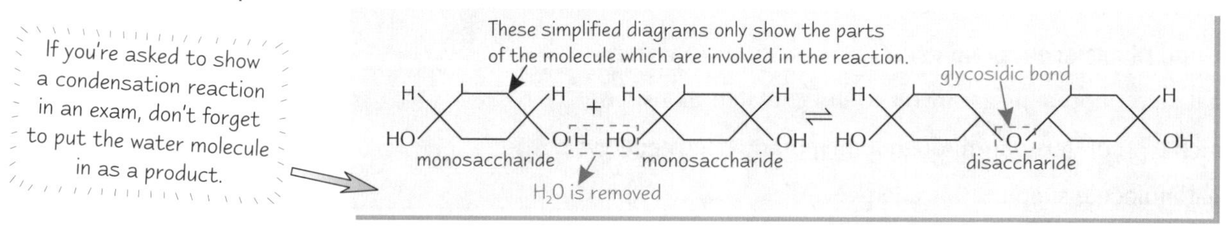

Hydrolysis Breaks Sugars Apart

When sugars are separated, the condensation reaction goes into **reverse**. This is called a **hydrolysis reaction** — a water molecule reacts with the glycosidic bond and breaks it apart.

disaccharide ⇌ monosaccharide + monosaccharide

H_2O reacts with glycosidic bond

It's all in the name — "hydro" is to do with water, and "lysis" means breaking up.

Condensation and hydrolysis reactions are dead important in biology. **Proteins and lipids** are put together and broken up by them as well. So you definitely need to understand how they work.

Carbohydrates

Polysaccharides are Loads of Sugars Joined Together

Polysaccharides are **polymers** — molecules which are made up of **loads of monomers** (**sugar molecules** in this case) stuck together. The ones you need to know about are:

1) **starch** — the main storage material in plants;
2) **glycogen** — the main storage material in animals;
3) **cellulose** — the major component of cell walls in plants.

Examiners like to ask about the link between the structures of polysaccharides and their functions.

① **Starch** is made up of **two** other polysaccharides of **alpha-glucose**:

- **Amylose** is a long, **unbranched chain** of alpha-glucose. The angles of the glycosidic bonds give it a **coiled structure**, almost like a cylinder. Its **compact**, coiled structure makes it really **good for storage**.
- **Amylopectin** is a long, **branched chain** of alpha-glucose. Its **side branches** make it particularly good for the storage of glucose — the enzymes that break down the molecule can get at the glycosidic bonds easily, to break them and release the glucose.

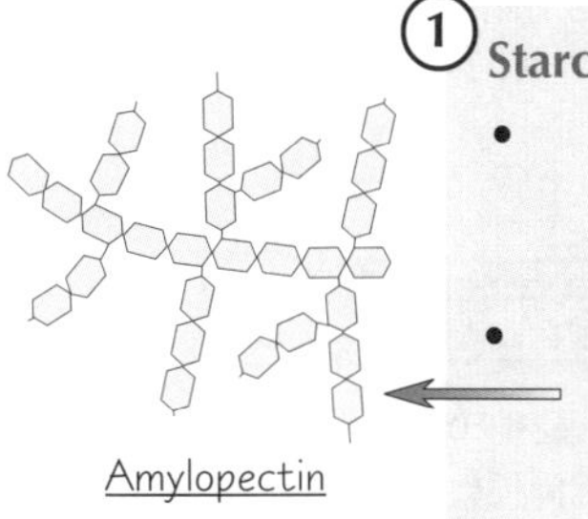
Amylopectin

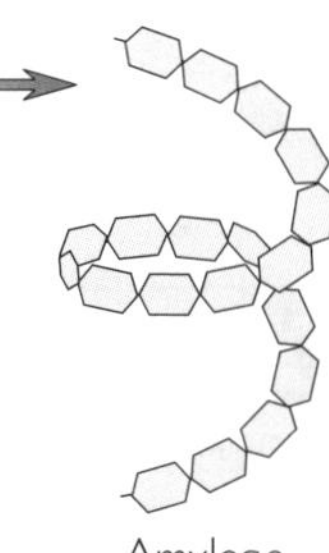
Amylose

② **Glycogen** is a polysaccharide of **alpha-glucose**. Its structure is very similar to amylopectin, except that it has **loads** more **side branches** coming off it. It's a very **compact** molecule found in animal liver and muscle cells. Loads of branches mean that stored glucose can be released quickly, which is **important for energy release** in animals.

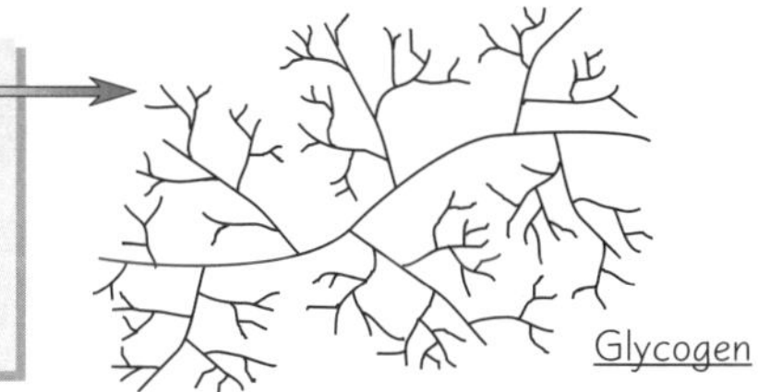
Glycogen

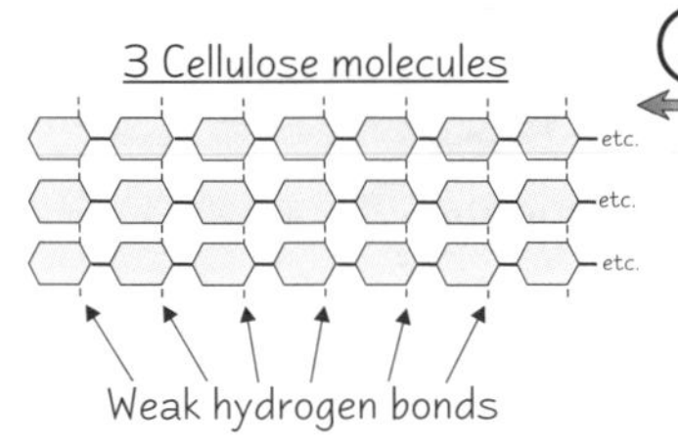

③ **Cellulose** is made of long, unbranched chains of **beta-glucose**. The bonds between beta sugars are **straight**, so the chains are straight. The chains are linked together by weak **hydrogen bonds** to form strong fibres called **microfibrils**. The strong fibres mean cellulose can provide **structural support** for cells. Another feature is that the **enzymes** that break the glycosidic bonds in starch can't reach the glycosidic bonds in cellulose, so those enzymes **can't break down cellulose**.

Practice Questions

Q1 What makes alpha glucose different from beta glucose?

Q2 What is the name given to the type of bond that holds sugar molecules together?

Q3 Explain the term "condensation reaction".

Q4 Name the two different types of molecule that are combined together in a starch molecule.

Q5 Name three polysaccharides and give the function of each one.

Q6 Cellulose is made from beta glucose. How does this help with its function as a structural polysaccharide?

Exam Questions

Q1 Describe how glycosidic bonds in carbohydrates are formed and broken in living organisms. [7 marks]

Q2 Compare and contrast the structures of glycogen and cellulose, showing how each molecule's structure is linked to its function. [10 marks]

Who's a pretty polysaccharide, then...

If you learn these basics it makes it easier to learn some of the more complicated stuff later on — 'cos carbohydrates crop up all over the place in biology. Remember that condensation and hydrolysis reactions are the reverse of each other — and don't forget that starch is composed of two polysaccharides. So many reminders, so little space ...

Proteins

There are hundreds of different proteins — all of them contain carbon, hydrogen, oxygen and nitrogen. They are the most abundant organic molecules in cells, making up 50% or more of a cell's dry mass — now that's just plain greedy.

Proteins are Made from Long Chains of Amino Acids

All proteins are made up of amino acids joined together. All amino acids have a **carboxyl group** (-COOH) and an **amino group** ($-NH_2$) attached to a carbon atom.

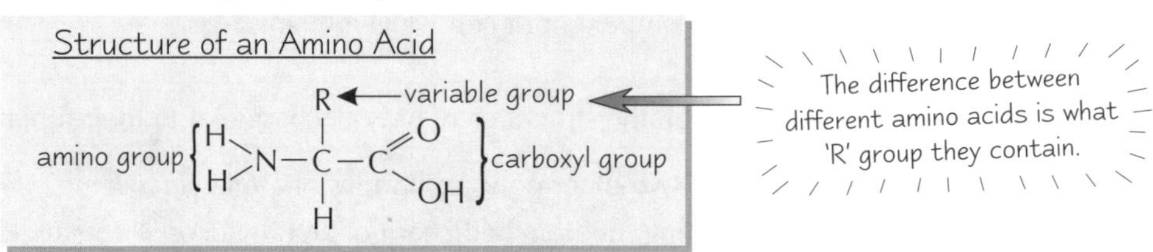

Proteins are Formed by...you guessed it...Condensation Reactions

Just like carbohydrates and lipids, the parts of a protein are put together by **condensation** reactions and broken apart by **hydrolysis** reactions. The bonds that are formed between amino acids are called **peptide bonds**.

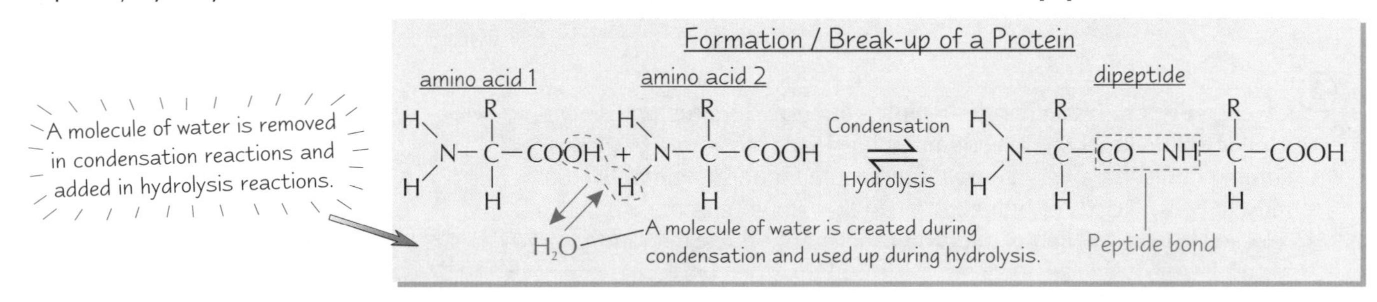

Proteins have up to Four Structures

Proteins are **big, complicated** molecules. They're easier to explain if you describe their structure in four 'levels'. These levels are called the protein's **primary, secondary, tertiary** and **quaternary** structures.

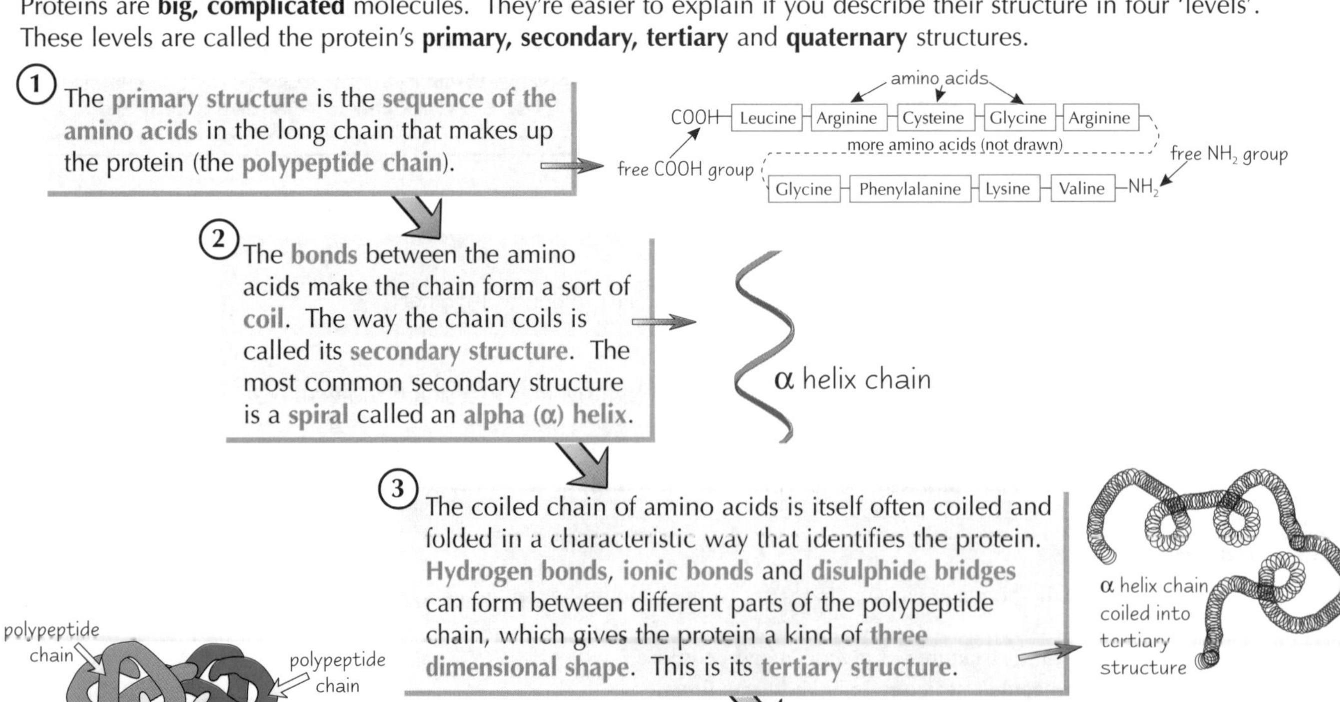

1. The **primary structure** is the **sequence of the amino acids** in the long chain that makes up the protein (the **polypeptide chain**).

2. The **bonds** between the amino acids make the chain form a sort of **coil**. The way the chain coils is called its **secondary structure**. The most common secondary structure is a **spiral** called an **alpha (α) helix**.

3. The coiled chain of amino acids is itself often coiled and folded in a characteristic way that identifies the protein. **Hydrogen bonds, ionic bonds** and **disulphide bridges** can form between different parts of the polypeptide chain, which gives the protein a kind of **three dimensional shape**. This is its **tertiary structure**.

4. Finally, some proteins are made of **several different polypeptide chains** held together by **various bonds**. The **quaternary structure** is the way these different parts are assembled together.

Proteins

Proteins are either Fibrous or Globular

Proteins can be separated into two groups — **fibrous proteins** and **globular proteins**. Each group has particular characteristics:

1) **Fibrous proteins** — these are made up of long, **insoluble polypeptide chains** which, in its tertiary structure, are **tightly coiled** round each other to form a **strong, fibrous structure**. An example of a fibrous protein is **collagen**:

Collagen is a protein that forms **supportive tissue** in animals, so it needs to be strong.

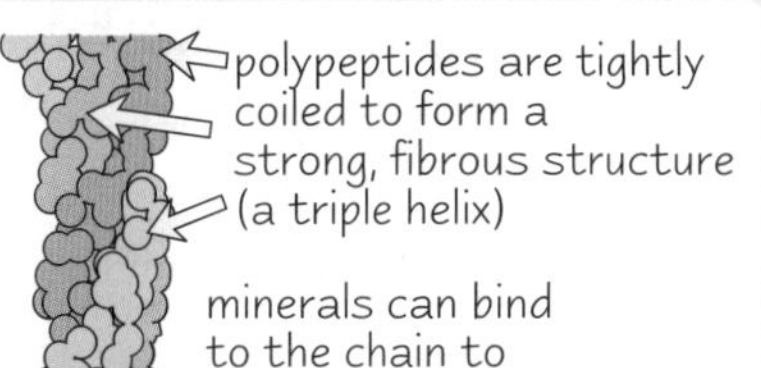

2) **Globular proteins** — these are made up of **polypeptide chains** that are **folded lots of times**, giving them a **complex tertiary structure**. Their round and compact tertiary shape means they are **soluble**, which makes them **easy to transport** round in the blood. Examples are **haemoglobin** and **insulin** — you need to know how their globular structure makes them good at their jobs:

Haemoglobin is a **globular protein** that absorbs oxygen (see diagram on p.12). Its structure is curled up, so **hydrophilic** ('water-attracting') side chains are on the **outside** of the molecule and **hydrophobic** ('water-repelling') side chains face **inwards**. This makes it soluble in water and good for transport in the blood.

Insulin is a hormone that reduces blood glucose levels.

it's a 'globular protein' — so it's good for transport

disulphide bonds hold the molecule in shape

it's a small molecule — so it's easily transported and absorbed by cells

Practice Questions

Q1 Name the four elements that are found in all proteins.

Q2 Draw the general structure of an amino acid.

Q3 What reaction joins protein molecules together? What reaction breaks them apart?

Q4 What is the name given to the bond that holds amino acids together in proteins?

Q5 Explain what a protein's a) primary and b) tertiary structure is.

Q6 Give three features of globular proteins.

Exam Questions

Q1 Describe the structure of a protein, explaining the terms primary, secondary, tertiary and quaternary structure. No details are required of the chemical nature of the bonds. [5 marks]

Q2 Describe the structure of the insulin molecule, and explain how this structure relates to its function in the body. [6 marks]

The name's Bond — Peptide Bond...

Quite a lot to learn on these pages — proteins are annoyingly complicated. Not happy with one, or even two, structures — they've got four of the things, and you need to learn 'em all. Condensation and hydrolysis reactions are back by popular demand and you need to learn examples of how globular proteins are adapted to their functions too.

Lipids

Lipids are fats, oils and waxes — they are all made up of carbon, hydrogen and oxygen, and they're all insoluble in water. Ever seen a candle dissolve in water? No — exactly.

Lipids** are Fats, Oils and Waxes — they're **Useful

1) Lipids contain a lot of **energy per gram**, so they make useful **medium** or **long-term energy stores**. But they can't be broken down very quickly, so organisms use carbohydrates for **short-term storage**.
2) Lipids stored under the skin in **mammals** act as **insulation**. Skin loses heat from blood vessels, but the fatty tissue under the skin doesn't have an extensive blood supply, so it conserves heat.

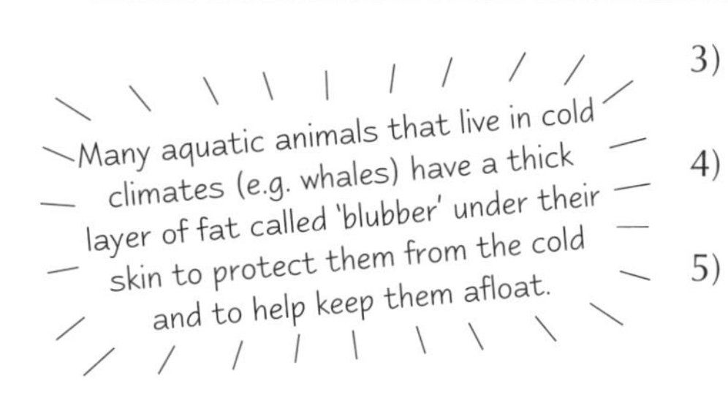

3) In **marine mammals** (e.g. whales, seals) lipids provide **buoyancy**, because the density of lipids is lower than that of muscle and bone.
4) Lipids under the skin and around the internal organs also provide **physical protection**, acting as a **cushion** against blows.
5) Lipids can act as a **waterproofing** layer — for example in the **waxy cuticle** on the surface of leaves and in the **exoskeleton** of insects. Lipids don't mix well with water, so water can't get through a lipid layer very easily.

*Most Fats and Oils are **Triglycerides***

Most lipids are composed of compounds called triglycerides. Triglycerides are composed of one molecule of **glycerol** with **three fatty acids** attached to it.

Fatty acid molecules have long 'tails' made of **hydrocarbons**. The tails are '**hydrophobic**' (they repel water molecules). These tails make lipids insoluble in water. When put in water, fat and oil molecules **clump together** in globules to reduce the surface area in contact with water.

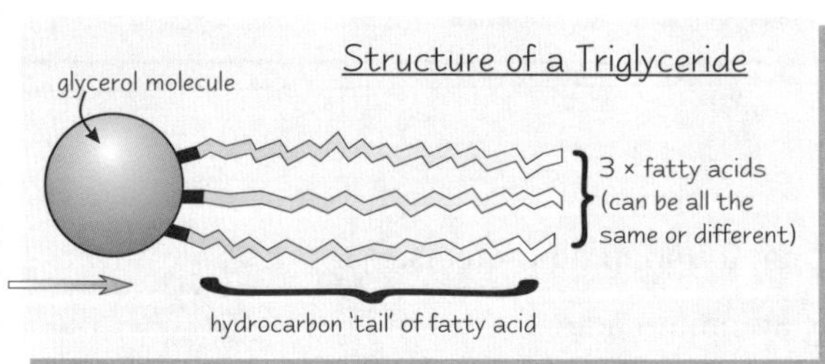

All **fatty acids** consist of the same basic structure, but the **hydrocarbon tail varies**. The tail is shown in the diagram with the letter 'R'.

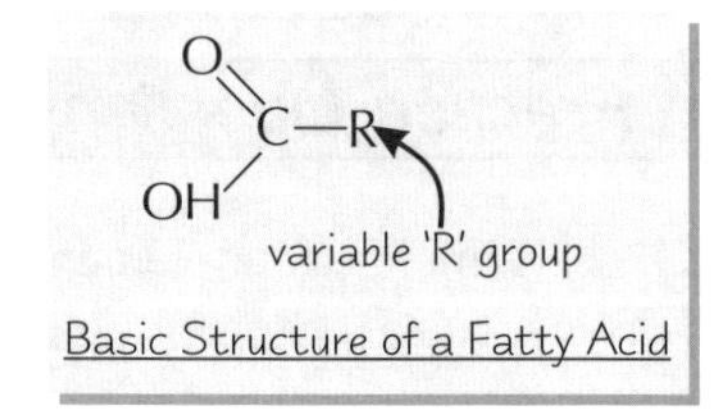

*Triglycerides are **Formed** by **Condensation Reactions***

Like carbohydrates, lipids are formed by **condensation reactions** and broken up by **hydrolysis reactions**.

The diagram below shows a **fatty acid** joining to a **glycerol molecule**, forming an **ester bond**. A molecule of water is also formed — it's a **condensation reaction**. This process happens twice more, to form a **triglyceride**. The **reverse** happens in **hydrolysis** — a molecule of water is added to each ester bond to break it apart, and the triglyceride splits up into three fatty acids and one glycerol molecule.

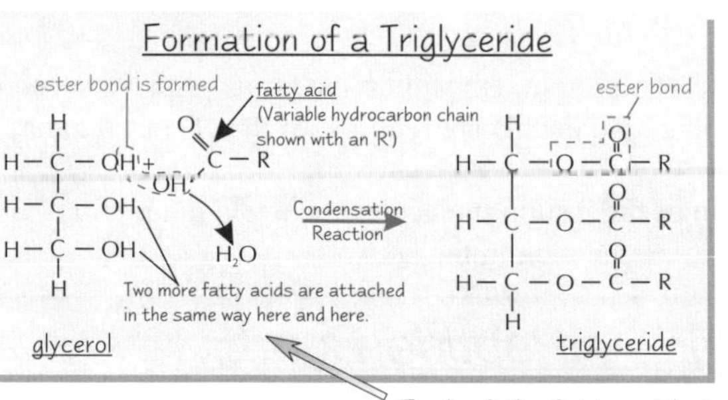

Each of the fatty acids in a triglyceride is attached to the glycerol molecule by an ester bond.

Lipids

Lipids can be Saturated or Unsaturated

There are two kinds of lipids — **saturated** lipids and **unsaturated** lipids. **Saturated** lipids are mainly **animal fats** and **unsaturated** lipids are found mostly in **plants** (unsaturated lipids are called **oils**). The difference between these two types of lipids is in their **hydrocarbon tails**.

1) Saturated fats **don't** have any **double bonds** between their carbon atoms — every bond has a **hydrogen** atom attached. The lipid is 'saturated' with hydrogen.
2) Unsaturated fats **do** have double bonds between carbon atoms. If they have **two or more** of them, the fat is called **polyunsaturated** fat.

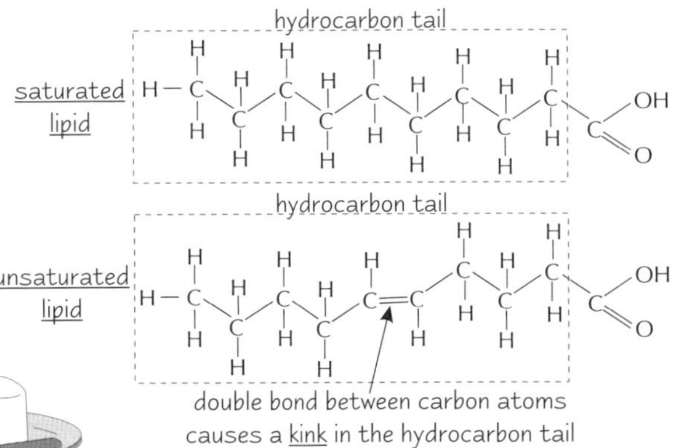

Unsaturated fats melt at lower temperatures than saturated ones. When used in margarine or butter spreads, it makes them easier to use 'straight from the fridge'.

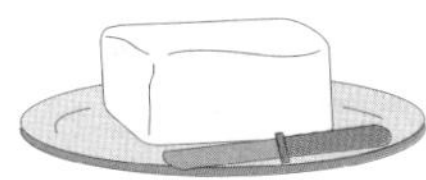

Phospholipids are a Special Type of Lipid

The lipids found in **cell membranes** aren't triglycerides — they're **phospholipids**. The difference is small but important:

1) In phospholipids, a **phosphate group** replaces one of the fatty acid molecules.
2) The phosphate group is **ionised**, which makes it **attract water** molecules.
3) So part of the phospholipid molecule is **hydrophilic** (attracts water) while the rest (the fatty acid tails) is **hydrophobic** (repels water). This is important in the cell membrane (see p.6).

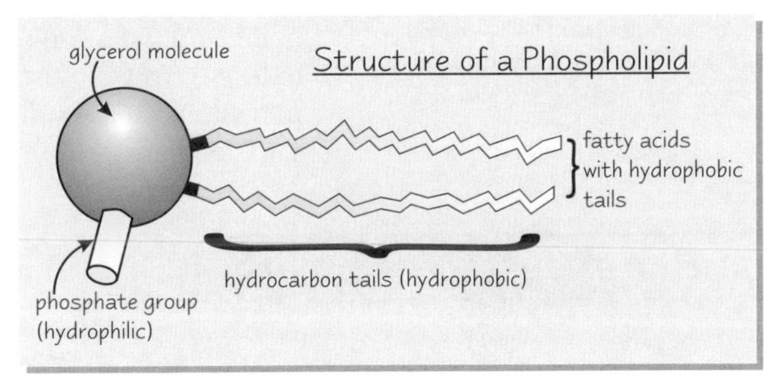

Practice Questions

Q1 Why is it wrong to call lipids 'fats'?

Q2 Why are lipids insoluble in water?

Q3 What's the name given to the type of bond that joins fatty acids to glycerol in a lipid molecule?

Q4 Explain the difference between a triglyceride and a phospholipid.

Exam Questions

Q1 Describe the chemical reactions involved in the assembly and breakdown of triglycerides in living organisms. [8 marks]

Q2 Describe the differences between a triglyceride and a phospholipid, and explain how these differences affect the properties of the molecule. [8 marks]

What did the seal say to the unhappy whale? — Quit blubbering...

Truly awful joke — I hang my head in shame. You don't get far in life without extensive lard knowledge, so learn all the details on this page good and proper. Lipids pop up in other sections, so make sure you know the basics about how their structure gives them some quite groovy properties. Right, all this lipid talk is making me hungry — chips time...

Biochemical Tests for Molecules

Here's a bit of light relief for you — two pages all about how you test for different food groups. There's nothing very complicated, you just need to remember a few chemical names and some colour changes.

Use the **Benedict's Test** to test for **Reducing Sugars**

The Benedict's test identifies **reducing sugars**. These are sugars that can donate electrons to other molecules — they include **all monosaccharides** and **some disaccharides**, e.g. maltose. When added to reducing sugars and heated, the **blue Benedict's reagent** gradually turns **brick red** due to the formation of a **red precipitate**.

The colour changes from:

blue — **green** — **yellow** — **orange** — **brick red**.

The higher the concentration of reducing sugar, the further the colour change goes — you can use this to **compare** the amount of reducing sugar in different solutions. A more accurate way of doing this is to **filter** the solution and **weigh the precipitate**.

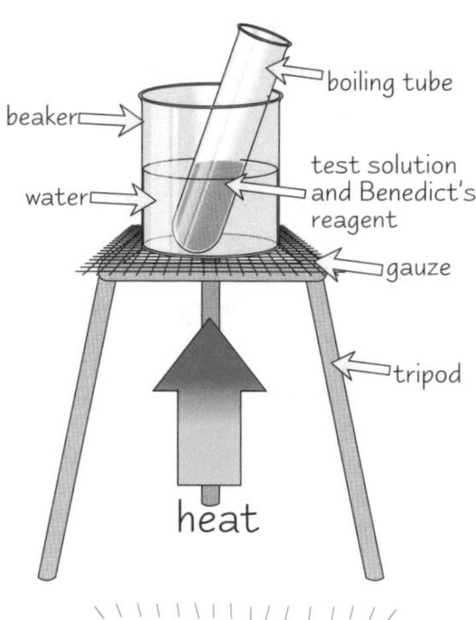

To test for **non-reducing sugars** like sucrose, which is a disaccharide (two monosaccharides joined together), you first have to break them down chemically into monosaccharides. You do this by boiling the test solution with **dilute hydrochloric acid** and then neutralising it with sodium hydrogen carbonate before doing the Benedict's test.

See p.10 for more on monosaccharides and disaccharides.

Use the **Iodine Test** to test for **Starch**

Make sure you always talk about iodine in potassium iodide solution, not just iodine.

In this test, you don't have to make a **solution** from the substance you want to test — you can use **solids** too. Dead easy — just add **iodine dissolved in potassium iodide solution** to the test sample. If there's starch present, the sample changes from **browny-orange** to a dark, **blue-black** colour.

Use the **Biuret Test** for **Proteins**

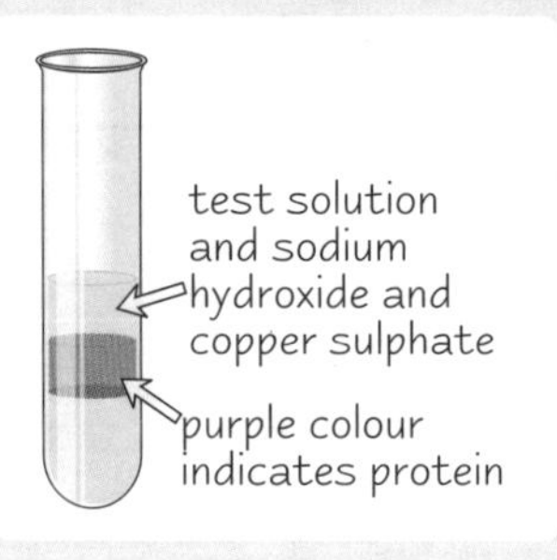

There are **two stages** to this test.

1) The test solution needs to be **alkaline**, so first you add a few drops of **2M sodium hydroxide**.
2) Then you add some **0.5% copper (II) sulphate solution**. If a **purple layer** forms, there's protein in it. If it stays **blue**, there isn't. The colours are pale, so you need to look carefully.

Use the **Emulsion Test** for **Lipids**

Shake the test substance with **ethanol** for about a minute, then pour the solution into water. Any lipid will show up as a **milky emulsion**. The more lipid there is, the more noticeable the milky colour will be.

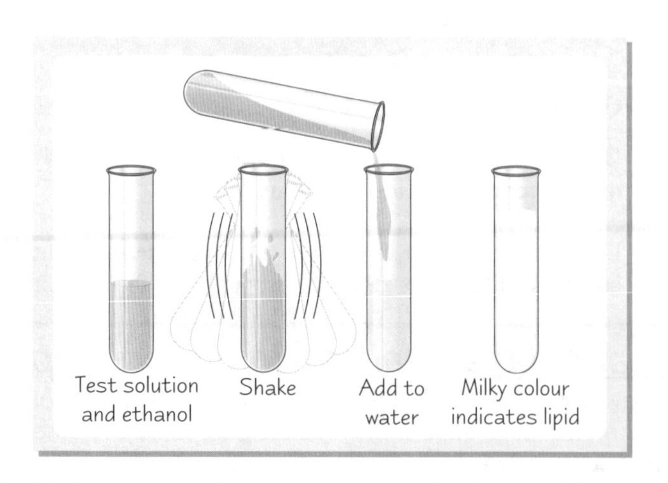

Chromatography

Chromatography Separates Out Molecules

If you have a mixture of biological chemicals in a sample that you want to test, you can separate them using the technique of **chromatography**.

1) You put a spot of the test solution onto a strip of special **chromatography paper**, then dip the end of the strip into a **solvent**.
2) As the solvent spreads up the paper, the different chemicals move with it, but at **different rates**, so they separate out.
3) You can identify what the chemicals are, using their **R_f values** (see diagram below).

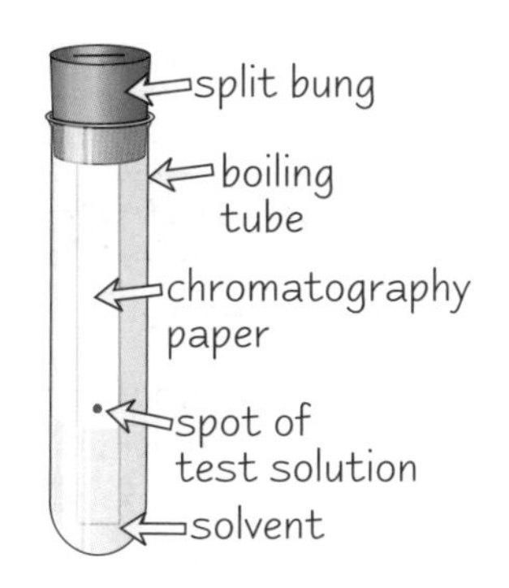

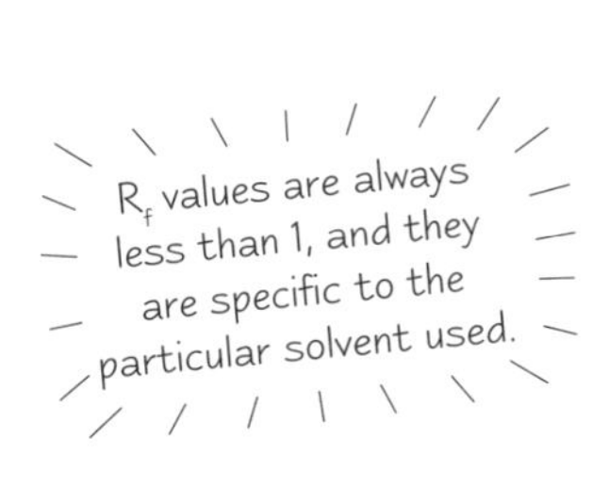

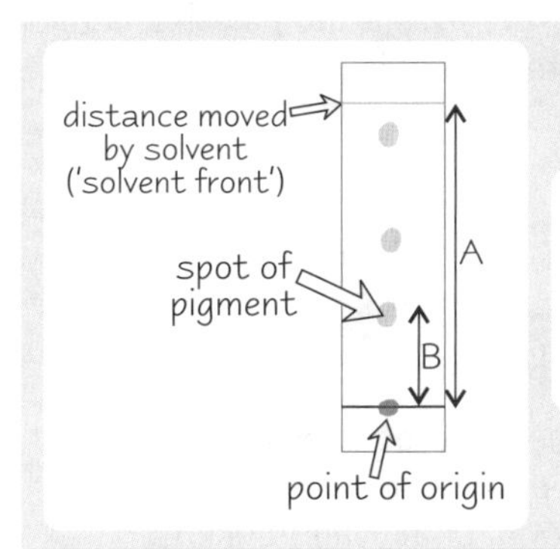

R_f value of pigment $= \frac{B}{A}$

$= \frac{\text{distance travelled by spot}}{\text{distance travelled by solvent}}$

Sometimes, the solvent doesn't completely separate out all the chemicals. In this case you need to use **two-way chromatography**, which uses a second solvent to complete the separation.

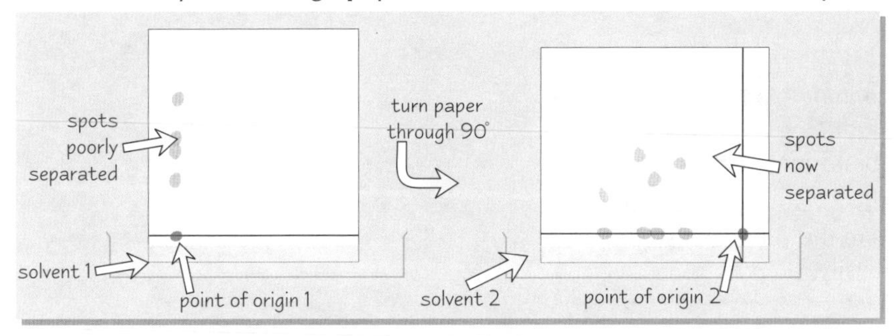

Some chemicals, like amino acids, aren't coloured, which makes it hard to identify them using chromatography. There are various ways you can colour them, though (e.g. by adding the chemical, ninhydrin).

Practice Questions

Q1 How can you work out the different concentrations of reducing sugars in two solutions?

Q2 Describe how you would test a solution for starch. What result would you expect if:
a) starch was present? b) starch was not present?

Q3 How is an "R_f value" of a chemical calculated?

Q4 When would you use "two-way" chromatography?

Exam Questions

Q1 You are given an unknown solution to test for different biochemical groups.
Describe the tests you would carry out and how you would analyse the results. [14 marks]

Q2 Describe how you would separate and identify the different pigments in a leaf extract by means of chromatography. [7 marks]

The Anger Test — Annoy the test subject. If it goes red, anger is present...

The days of GCSEs might have gone forever, but with this page you can almost feel like you're back there in the mists of time, when biology was easy and you fancied someone out of S-Club 7. Aah. Well, you'd better make the most of it and get these tests learnt — 'cos things get trickier than a world-class magician later on...

Action of Enzymes

*Enzymes crop up loads in biology — they're really useful 'cos they make reactions work more quickly. So, whether you feel the need for some speed or not, read on — because you **really** need to know this basic stuff about enzymes.*

Enzymes are Biological Catalysts

Enzymes speed up chemical reactions by acting as **biological catalysts**.

1) They catalyse every **metabolic reaction** in the bodies of living organisms. Even your **phenotype** (physical appearance) is down to enzymes that catalyse the reactions that cause growth and development.
2) Enzymes are **globular proteins** (see p.13) although some have **non-protein components** too.
3) Every enzyme has an area called its **active site**. This is the part that connects the enzyme to the substance it interacts with, which is called the **substrate**.

A catalyst is a substance that speeds up a chemical reaction without being used up in the reaction itself.

Enzymes Reduce Activation Energy

In a chemical reaction, a certain amount of energy needs to be supplied to the chemicals before the reaction will start. This is called the **activation energy** — it's often provided as **heat**. Enzymes **reduce** the amount of activation energy that's needed, often making reactions happen at a **lower temperature** than they could without an enzyme. This **speeds** up the **rate of reaction**.

When a substrate fits into the enzyme's active site it forms an **enzyme-substrate complex**:

1) If two substrate molecules need to be **joined**, attaching to the enzyme holds them **close together**, **reducing** any **repulsion** between the molecules so they can bond more easily.
2) If the enzyme is catalysing a **breakdown reaction**, fitting into the active site puts a **strain** on bonds in the substrate, so the substrate molecule **breaks up** more easily.

Graph Showing How Enzymes Speed up the Rate of Reaction

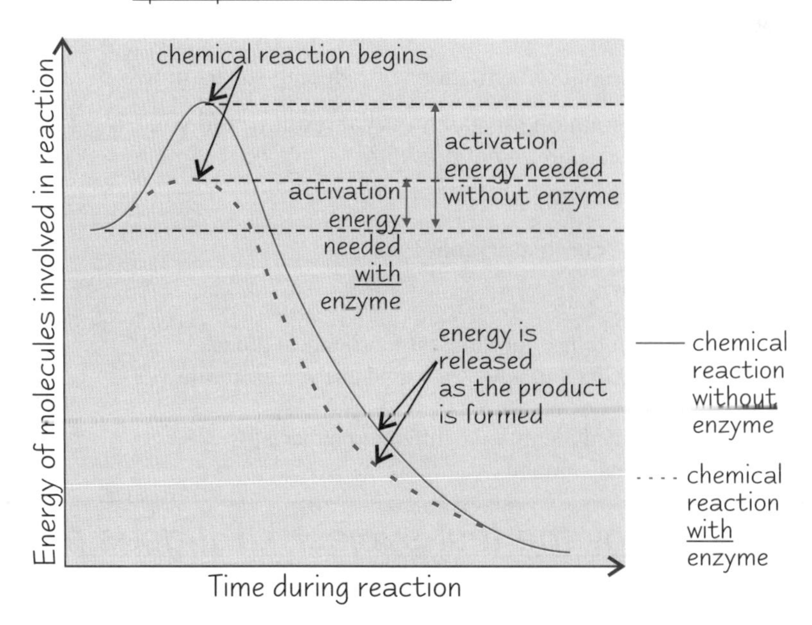

Action of Enzymes

The 'Lock and Key' Model is a Good Start...

Enzymes are a bit picky. They only work with **specific substrates** — usually only one. This is because, for the enzyme to work, the substrate has to **fit** into the **active site**. If the substrate's shape doesn't match the active site's shape, then the reaction won't be catalysed. This is called the '**lock and key**' model.

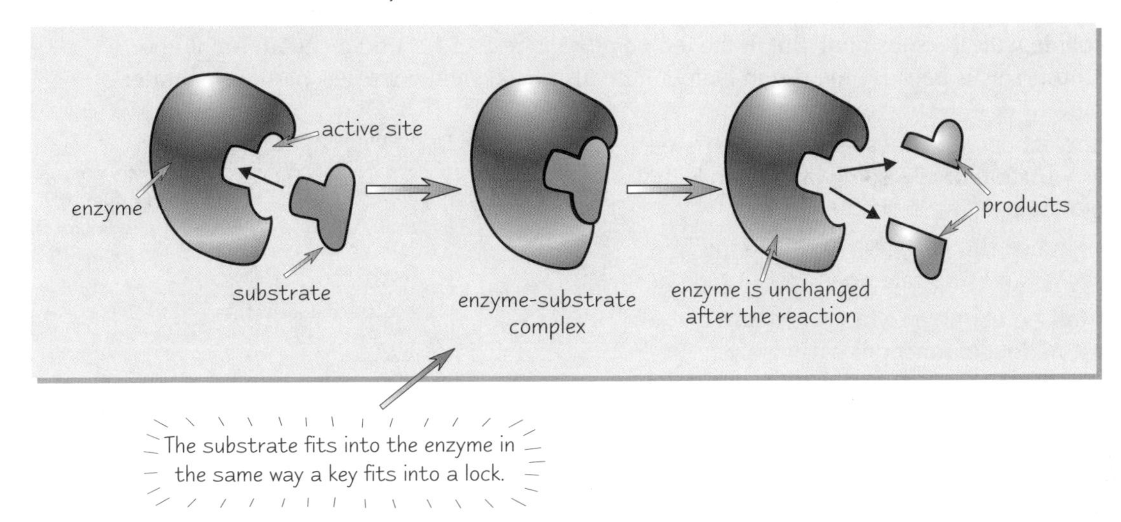

The substrate fits into the enzyme in the same way a key fits into a lock.

...but the 'Induced Fit' Model is a Better Theory

Scientists now believe that the lock and key model doesn't tell the whole story. The enzyme and substrate do have to fit together in the first place, but then it seems that the **enzyme-substrate complex changes shape** slightly to complete the fit. This **locks** the substrate even more tightly to the enzyme. This is called the '**induced fit**' model. It helps explain why enzymes are so **specific** and only bond to one particular substrate. The substrate doesn't only have to be the right shape to fit the active site, it has to make the active site **change shape** in the right way as well.

The 'Luminous Tights' model was popular in the 1980s but has since been found to be grossly inappropriate.

Practice Questions

Q1 Define the term "catalyst".

Q2 How is your "phenotype" affected by enzymes?

Q3 What is the name given to the amount of energy needed to start a reaction?

Q4 What is an "enzyme-substrate complex"?

Q5 Explain why enzymes are specific (i.e. only work with a single or a small group of substrates).

Exam Questions

Q1 Explain how enzymes act as "biological catalysts". [3 marks]

Q2 Explain the differences between the "lock and key" model of enzyme action, and the "induced fit" model. [7 marks]

But why is the enzyme-substrate complex?

OK, nothing too tricky here. The main thing to remember is that every enzyme has a specific shape, so it only works with specific substrates that fit the shape. The induced fit model is the new, trendy theory to explain this — the lock and key model is, like, ***so*** *last year. Everyone who's anyone knows that.*

Factors that Affect Enzyme Activity

Just when you thought you'd seen the last of enzymes, here they are again to brighten up your day one more time. This time it's all about the conditions that enzymes work best in.

Temperature has a Big Influence on Enzyme Activity

Like any chemical reaction, the rate of an enzyme-controlled reaction increases when the temperature's raised. More heat means more **kinetic energy**, so molecules move faster. This makes the enzyme more likely to **collide** with the substrate. But, if the temperature increases beyond a certain point, the **reaction stops**. This is because the rise in temperature also makes the enzyme's particles **vibrate**:

1) If the temperature goes above a certain level, this vibration **breaks** some of the **bonds** that hold the enzyme in shape.
2) The **active site changes shape** and the enzyme and substrate **no longer fit together**.
3) At this point, the enzyme is **denatured** — it no longer functions as a catalyst.

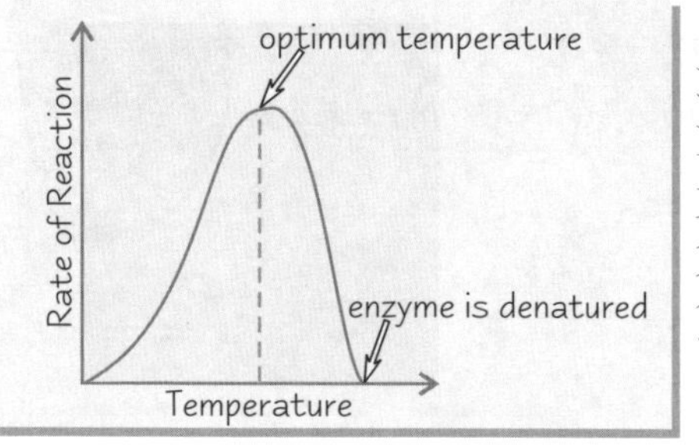

Every enzyme has an optimum temperature. In humans it's around 37°C but some enzymes, like those used in biological washing powders, can work well at 60°C.

pH Also Affects Enzyme Activity

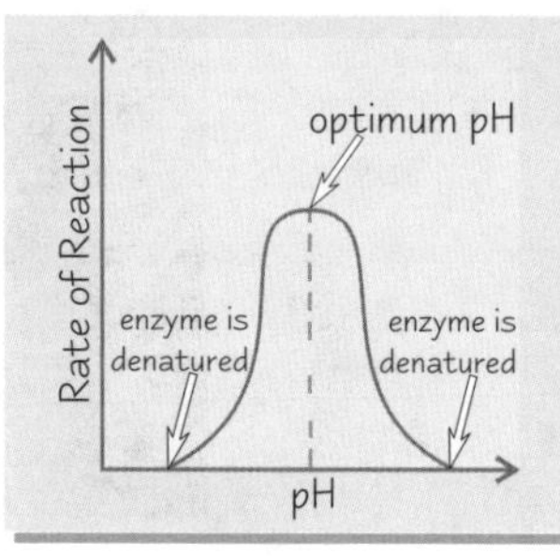

All enzymes have an **optimum pH value**. Most work best at neutral pH 7, but there are exceptions. **Pepsin**, for example, works best at acidic pH 2, which suits it to its role as a stomach enzyme. Above and below the optimum pH, the H+ and OH- ions found in acids and alkalis can mess up the **ionic bonds** that hold the enzyme's tertiary structure in place. This makes the active site change shape, so the enzyme is **denatured**.

Enzyme Concentration Affects the Rate of Reaction

1) The **more enzyme molecules** there are in a solution, the more likely a substrate molecule is to **collide** with one. So increasing the concentration of the enzyme increases the rate of reaction.
2) But if the amount of substrate is limited, there comes a point when there's more than enough enzyme to deal with all the available substrate, so adding more enzyme has **no further effect**.

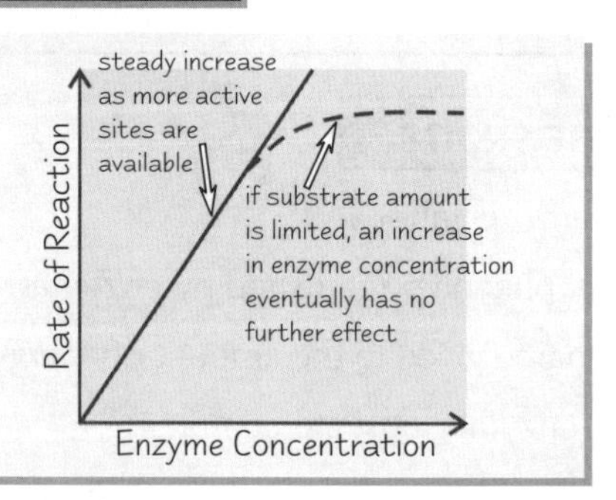

Substrate Concentration Affects the Rate of Reaction Up To a Point

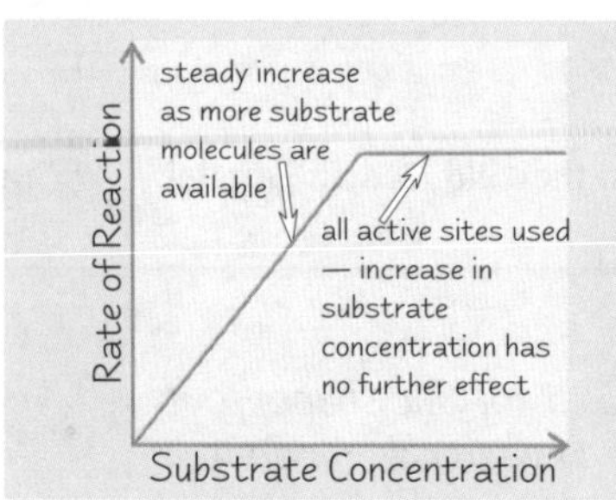

Substrate concentration affects the rate of reaction up to a certain point. The higher the substrate concentration, the faster the reaction, but only up until a **'saturation' point**. After that, there's so many substrate molecules that the enzymes have about as much as they can cope with, and adding more **makes no difference**.

Factors that Affect Enzyme Activity

Enzyme Activity can be *Inhibited*

Enzyme activity can be prevented by **enzyme inhibitors** — molecules that **bind to the enzyme** that they inhibit. Inhibition can be **competitive** (active site directed) or **non-competitive** (non-active site directed).

1) **Competitive inhibitors** have a **similar shape to the substrate**. They compete with the substrate to bond to the active site, but no reaction follows. Instead they **block** the active site, so **no substrate** can **fit** in it. How much inhibition happens depends on the **relative concentrations** of inhibitor and substrate — if there's a lot of the inhibitor, it'll take up all the active sites and stop any substrate from getting to the enzyme.

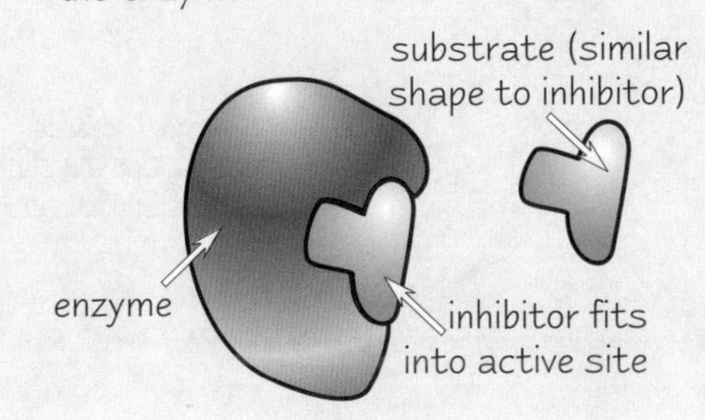

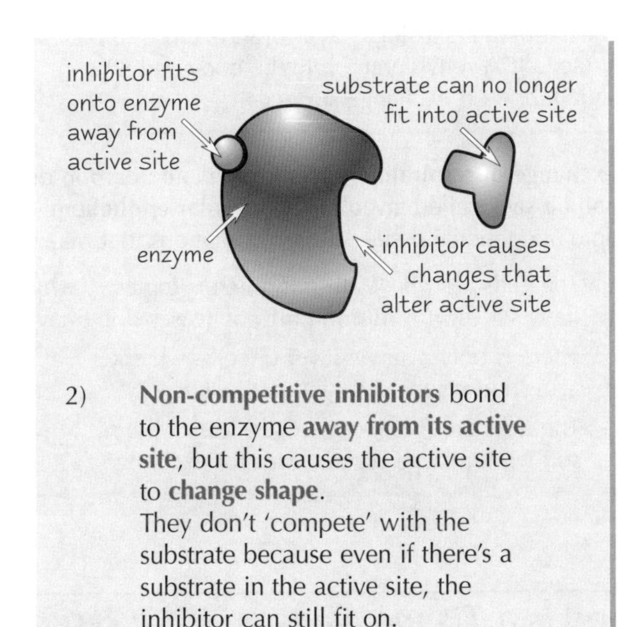

2) **Non-competitive inhibitors** bond to the enzyme **away from its active site**, but this causes the active site to **change shape**.
They don't 'compete' with the substrate because even if there's a substrate in the active site, the inhibitor can still fit on.

Practice Questions

Q1 Why do high temperatures denature enzymes?

Q2 What kind of bonds in enzymes are affected by the wrong pH?

Q3 Why does increasing enzyme concentration speed up the rate of reaction?

Q4 Explain why increasing the concentration of a substrate doesn't always increase the rate of reaction.

Q5 What is the difference between a competitive and a non-competitive enzyme inhibitor?

Exam Questions

Q1 When doing an experiment on enzymes, explain why it is necessary to control the temperature and pH of the solutions involved. [8 marks]

Q2 When a small amount of chemical X is added to a mixture of an enzyme and its substrate, the formation of reaction products is reduced. Increasing the amount of X in the solution causes further reduction in products. State, with reasons, the likely nature of chemical X. [4 marks]

Don't be shy — lose your inhibitions and learn these pages...

It's not easy being an enzyme. They're just trying to get on with their jobs, but the whole world seems to be against them sometimes. High temperature, wrong pH, inhibitors — they're all out to get them. Sad though it is, make sure you know every word. Learn how different factors affect enzyme activity, and be able to describe the different types of inhibitors.

Tissues

Tissues aren't just something that you blow a load of snot into. No, no, no — they're also groups of similar cells that are organised together to do specific functions. Blood is a tissue, believe it or not. Bless me.

Similar Cells are Organised into Tissues

A single-celled organism performs all its life functions in its one cell. **Multicellular organisms** (like us) are more complicated — different cells do different jobs, so cells have to be **organised** into different groups. Similar cells are grouped together into **tissues**.

You need to know about how **epithelial tissue** is adapted for its function:

Epithelial tissue is a **single layer** of **flat cells** lining a surface. It's pretty common in the body and is usually found on **exchange surfaces**.

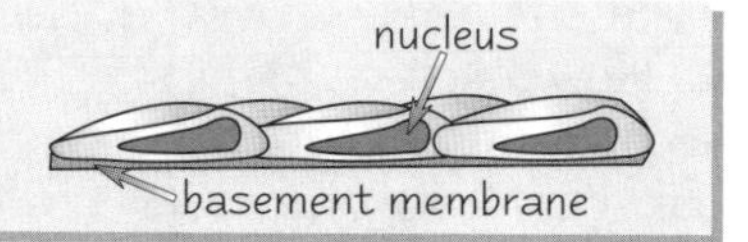

Epithelium means a tissue that forms a covering or a lining.

The **exchange of respiratory gases** (oxygen and carbon dioxide) happens in the **lungs**, in small air sacs called **alveoli**. The **alveolar epithelium**, that lines the alveoli, is an example of epithelial tissue. It has adaptations that make it efficient at gas exchange:

1) The cells are **thin**, with **not much cytoplasm**, which means there's a **short diffusion pathway** (gases don't have far to travel).
2) There is only a **single layer** of cells, which also shortens the diffusion pathway.
3) The cells are **flat** and there are **lots of them**, so there's a **large surface area** for exchange.

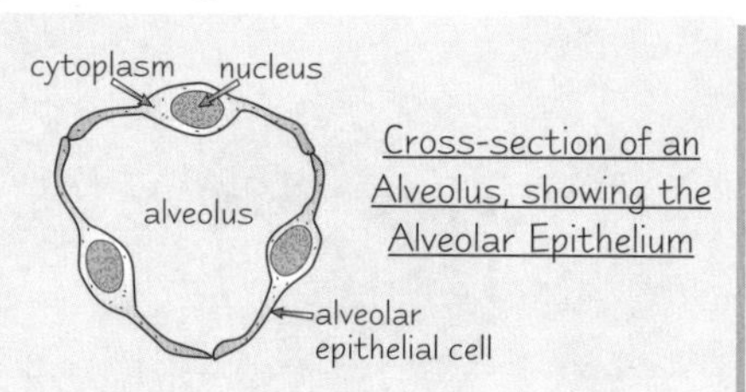

Cross-section of an Alveolus, showing the Alveolar Epithelium

Blood is a Tissue made up of Different Cells

Tissues aren't always made up of one type of cell. Some tissues include different types of cell working together. **Blood** is a **liquid tissue** that's composed of **different types of cell** suspended in a liquid called **plasma**. Plasma is mainly **water**, with various nutrients and gases dissolved in it.

1) **Red blood cells** are responsible for absorbing **oxygen** and transporting it round the body. They're made in the **bone marrow** and are very small.

Red Blood Cells

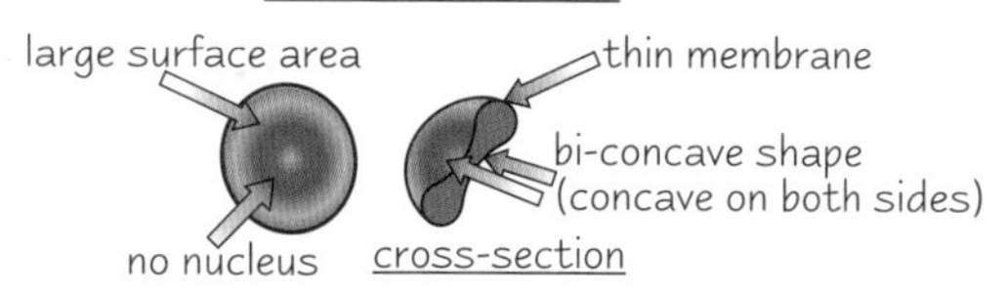

1) They have **no organelles** (including **no nucleus**) to leave more room for **haemoglobin**, which carries the oxygen.
2) They have a large surface area due to their **bi-concave disc** shape. This allows O_2 to diffuse quickly into and out of the cell.
3) They have an **elastic membrane**, which allows them to change shape to squeeze through the small blood capillaries, then spring back into normal shape when they re-enter veins.

2) **White blood cells** are larger than red blood cells but there are fewer of them in blood. They are responsible for fighting disease (see p.48). There are different types of white blood cell. The three examples you need to know about are **lymphocytes**, **monocytes**, and **granulocytes**:

Type of white blood cell	Diagram	Features
lymphocyte		• large nucleus
monocyte		• lobed nucleus
granulocyte		• lobed nucleus • granular cytoplasm

Surface Area to Volume Ratio

Smaller Animals have Higher *Surface Area : Volume Ratios*

Smaller animals have a **bigger surface area compared to their volume** than larger animals.
This can be hard to imagine, but you can prove it mathematically. Imagine these animals as cubes:

The hippo could be represented by a block with an area of
2 cm × 4 cm × 4 cm.

Its **volume** is 2 × 4 × 4 = **32 cm³**

Its **surface area** is 2 × 4 × 4 = 32 cm² (top and bottom surfaces of cube)
+ 4 × 2 × 4 = 32 cm² (four sides of the cube)

Total surface area = **64 cm²**

So the hippo has a **surface area : volume ratio** of 64 : 32 or **2 : 1**.

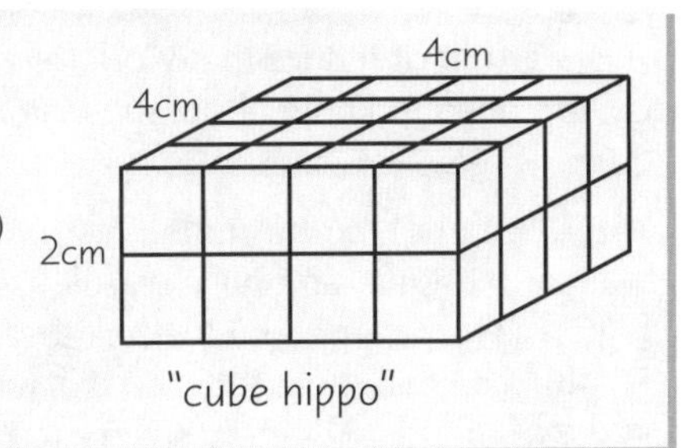

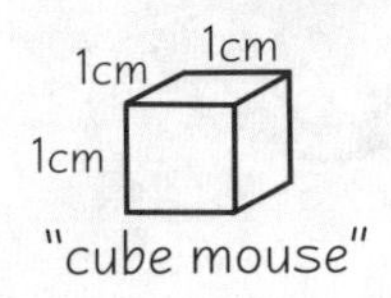

Compare this to a mouse cube measuring 1 cm × 1 cm × 1 cm

Its **volume** is 1 x 1 x 1 = **1 cm³**

Its **surface area** is 1 x 1 x 6 = **6 cm²**

So the mouse has a **surface area : volume ratio** of 6 : 1

The cube mouse's surface area is six times its volume. The cube hippo's surface area is only twice its volume.

Organisms Need to Exchange Materials *with the* Environment

Cells need **oxygen** (for aerobic respiration) and nutrients. They need to **excrete waste products** like CO_2.

1) Microscopic one-celled organisms have very **high** surface area : volume ratios, so they can exchange materials with the environment over their whole surface by diffusion.
2) Larger multicellular organisms have **lower** surface area : volume ratios. They need **organs** with **big surface areas** and **specialised cells** for exchange (e.g. lungs). Many have also developed specialised **transport systems** like blood, to carry gases, nutrients and wastes to and from inner cells.

Surface area is also important for **body temperature**. Animals release heat energy when their **cells respire** and lose it to the **environment**. So **small animals** with high S.A. : volume ratios, like mice, have to use lots of energy just **keeping warm**, while **big animals** with low S.A. : volume ratios, like elephants, are more likely to **overheat**.

Animals with low S.A. : volume ratios have evolved adaptations to increase surface area. For example elephants have big, flat ears to increase their surface area to help heat escape from the body quickly.

Practice Questions

Q1 What is the definition of a tissue?

Q2 What is the liquid part of blood called?

Q3 What do white blood cells do?

Q4 Why is surface area to volume ratio important in organisms?

Q5 Give an example of how an animal has adapted to increase its surface area to volume ratio.

Exam Questions

Q1 Explain how the structure of red blood cells helps them to transport oxygen efficiently. [3 marks]

Q2 Explain why a specialised gas exchange system is required in humans. [4 marks]

Cube animals indeed — it's all gone a bit Picasso...

Make sure you learn the features of alveolar epithelial tissue and blood tissue, and how they're adapted to their jobs — examiners love examples. Then move on to hippos and mice and all that malarkay. Hmm, if I were a blood cell I'd be a great lymphocyte warrior, instilling fear into the nuclei of pathogens everywhere. Aaanyway.

Organs and Blood Transport

For some people the word 'blood' is enough to make them cringe. Strange really, seeing as it's the substance that keeps us all alive. Anyway, these pages are about how it's transported round the body, and how it creates tissue fluid.

Tissues are Organised into Organs

An **organ** is a **group of tissues** that work together to perform a particular function. **Blood vessels** are examples of organs. You need to know about the features of three different blood vessels — **arteries**, **arterioles** and **veins**.

1) **Arteries** carry blood **from** the heart **to** the rest of the body. They're thick-walled, muscular and have elastic tissue in the walls to cope with the **high pressure** caused by the heartbeat. The inner lining (**endothelium**) is **folded** so that the artery can **expand** when the heartbeat causes a **surge of blood**. All arteries carry **oxygenated** blood except the **pulmonary arteries**, which take deoxygenated blood to the lungs.

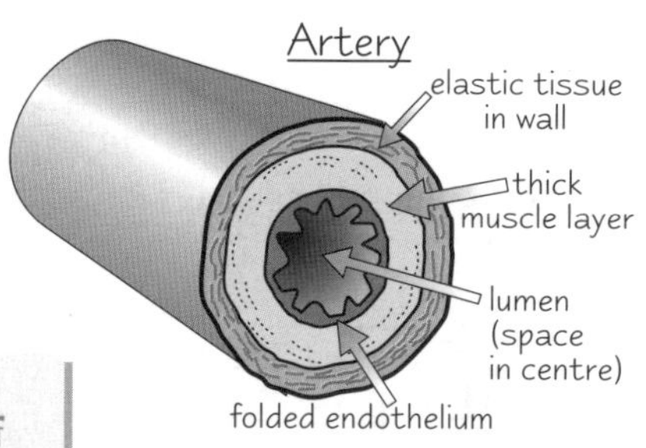

2) Arteries divide into smaller **arterioles** which form a network of vessels throughout the body. Blood is directed to different **areas of demand** in the body by muscles inside the arterioles contracting and restricting the blood flow or relaxing and allowing full blood flow.

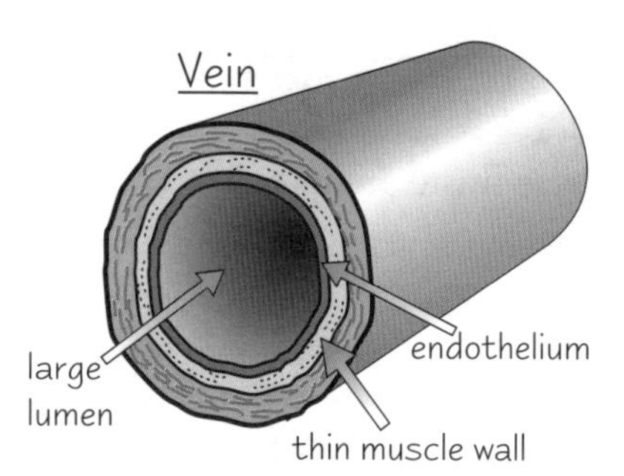

3) **Veins** take blood under **low pressure** back **to the heart**. They're **wider** than equivalent arteries, with very little elastic or muscle tissue. Veins contain **valves** to stop the blood flowing backwards. Blood flow through the veins is helped by contraction of the **body muscles** surrounding them. All veins carry **deoxygenated** blood (because oxygen has been used up by body cells), except for the **pulmonary veins**, which carry oxygenated blood to the heart from the lungs.

Mammals Have a Closed Double Circulation

Multicellular organisms, like **mammals**, have a **low surface area to volume ratio** (see p.23), so they need specialised **transport systems** to carry raw materials from specialised **exchange organs** to their body cells. In mammals this is the **closed circulatory system**, which uses **blood** to transport respiratory gases, products of digestion, metabolic wastes and hormones round the body.
The bulk movement of blood is called **mass flow**.

The mammalian closed circulatory system uses the **heart** to pump blood through **blood vessels** (like arteries, arterioles and veins) to reach different parts of the body. ⟶

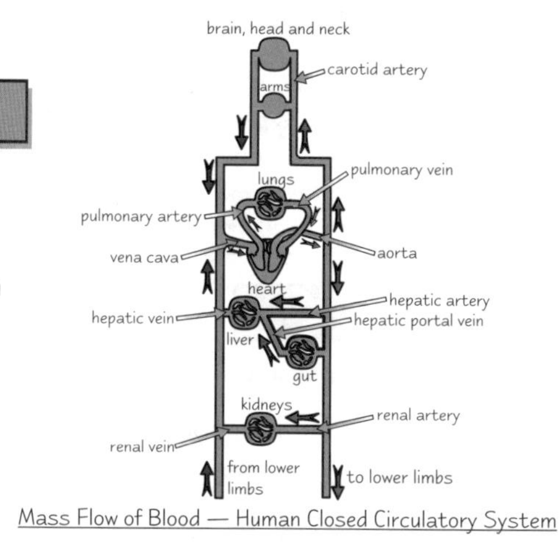

Mass Flow of Blood — Human Closed Circulatory System

Substances are Exchanged between Blood and Body Tissues at Capillaries

Arterioles branch into **capillaries**, which are the **smallest** of the blood vessels. Substances are **exchanged** between cells and capillaries, so they are adapted to carry out **efficient diffusion** of substances (e.g. glucose and oxygen):

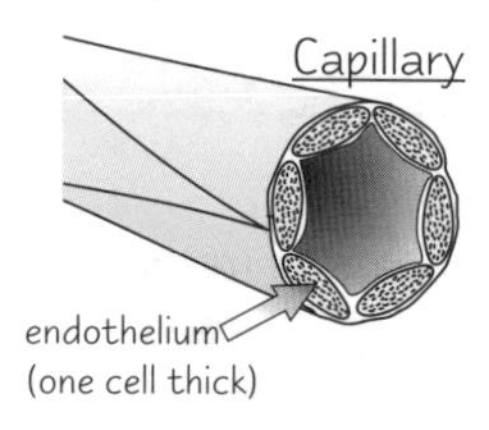

- They are always found very **near cells in exchange tissues**, so there's only a **short diffusion pathway**.
- Their walls are only **one cell thick**, which also shortens the diffusion pathway.
- There are a large number of capillaries, to **increase surface area** for exchange. Networks of capillaries in tissue are called **capillary beds**.

Tissue Fluid

Tissue Fluid is Formed from Blood Plasma

Tissue fluid surrounds all cells in tissues — providing them with the conditions they need to function. Tissue fluid is made from substances that leave the **plasma** from blood capillaries. These substances move out of blood capillaries due to a **pressure gradient**:

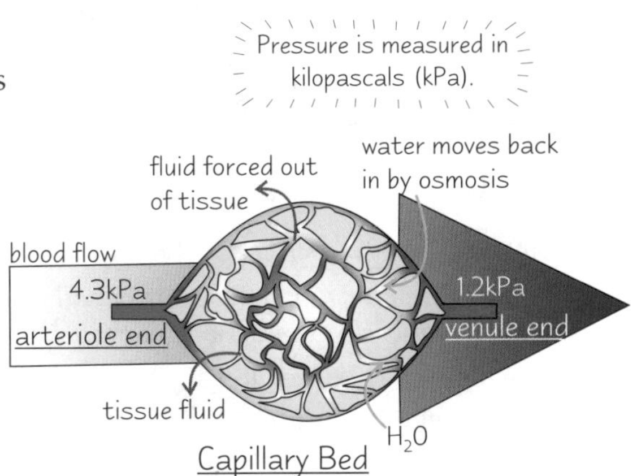

1) At the **arteriole end** of a capillary bed, pressure **inside** the capillaries is **greater** than pressure in the tissue fluid around cells. This difference in pressure forces fluid to **leave the capillaries** and enter tissue space.
2) As fluid leaves, pressure reduces in the capillaries — so the pressure's much lower at the **venule end** of the capillary bed.
3) Due to the fluid loss, the **water potential** at the **venule end** of the capillaries is **lower** than the water potential in the **tissue fluid** — so some **water re-enters** the capillaries from the tissue fluid at the venule end, by **osmosis**.

Unlike blood, tissue fluid **doesn't** contain **red blood cells** or **big proteins**, because they are **too large** to be pushed out through the capillary walls. It does contain smaller molecules, e.g. oxygen, glucose and mineral ions. Tissue fluid helps cells to get the oxygen and glucose they need, and to get rid of the CO_2 and waste they don't need.

Lymph is Formed from Excess Tissue Fluid

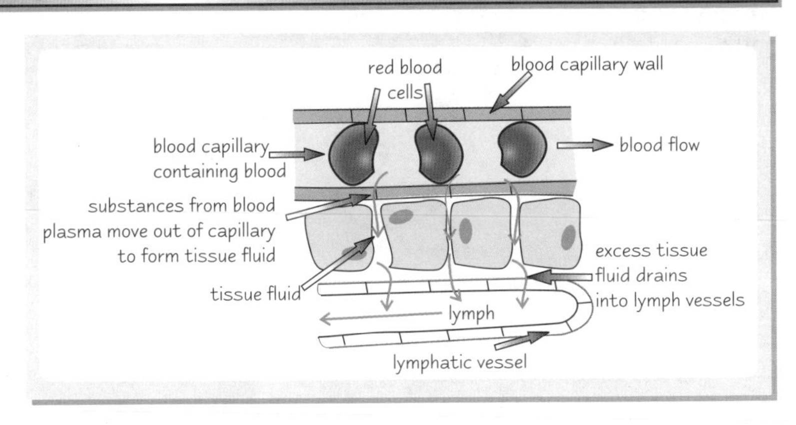

Lymph is a fluid that forms when excess tissue fluid drains into the lymphatic vessels, which lie close to blood capillaries. Lymph takes away waste products from the cells. It then travels through the **lymphatic system** and eventually enters blood plasma.

It's similar to tissue fluid, except it contains more fats, proteins and white blood cells, which it picks up at **lymph nodes** as it travels through the lymphatic system.

Practice Questions

Q1 Name three types of blood vessel.

Q2 Do arteries mainly carry oxygenated or deoxygenated blood?

Q3 Explain how tissue fluid is formed.

Q4 What makes lymph different from tissue fluid?

Exam Questions

Q1 Give two features of an artery's structure, and relate them to its function. [4 marks]

Q2 How does tissue fluid move into and out of blood capillaries? [5 marks]

If blood can handle transport this efficiently, the trains have no excuse…

Four hours I was waiting at Preston this weekend. Four hours! Anyway, you may have noticed that biologists are obsessed with the relationship between structure and function, so whenever you're learning the structure of something, make sure you know how this relates to its function. Like the veins, arteries and capillaries on these pages, for example.

Lungs and Ventilation

Quite a lot to learn on this page I'm afraid, but some of it is related to stuff that's described earlier on in the book (like exchange surfaces) so it should be familiar. Right, take a deep breath and off you go...

Lungs are Specialised Organs for Breathing

Mammals like humans exchange oxygen and carbon dioxide through their **lungs**. The lungs have special **features** that make them well-adapted to **breathing**:

Cartilage — rings of strong but bendy cartilage keep the **trachea** open.
Smooth muscle — round the bronchi and bronchioles. Involuntary muscle **contractions** narrow the airways.
Elastic fibres — between the alveoli. Stretch the lungs when we breathe in and recoil when we breathe out to help push air out.
Pleural membrane — protective lining on the lungs.
Intercostal muscles — they contract and relax to control ventilation.

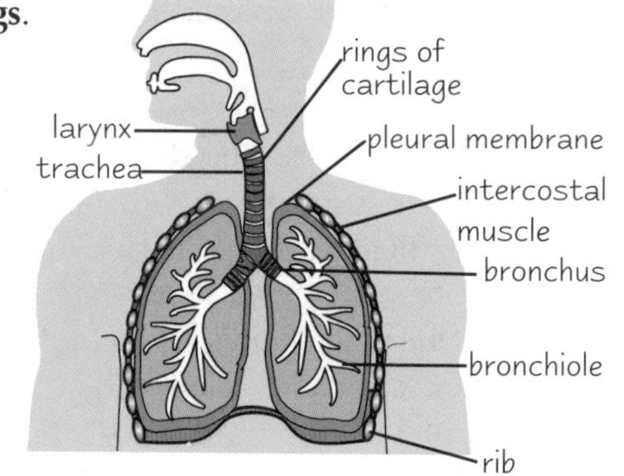

In Humans Gaseous Exchange Happens in the Alveoli

In the lungs, the bronchioles branch out into millions of microscopic air sacs called **alveoli**, which are responsible for gas exchange. They're so tiny, but so important — size ain't everything.

1) **O_2** diffuses **out of** alveoli, across the **alveolar epithelium** (the single layer of cells lining the alveoli) and the **capillary endothelium** (the single layer of cells of the capillary wall), and into **haemoglobin** in the **blood**.
2) **CO_2** diffuses **into** the alveoli from the blood, and is breathed out through the lungs, up the trachea, and out of the mouth and nose.

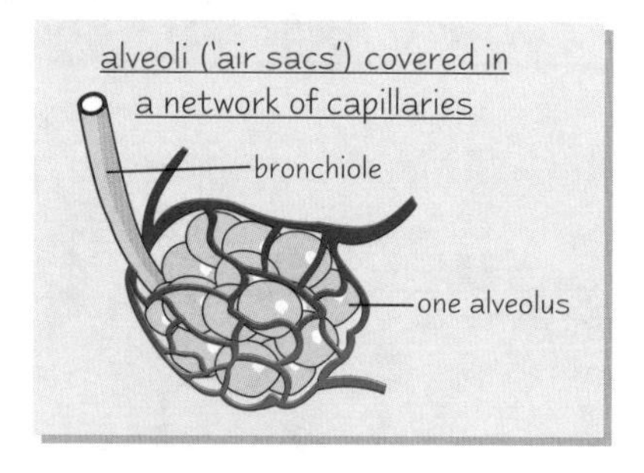

Fick's law (remember it from p.7?) shows the factors that determine rate of diffusion:

$$\text{rate of diffusion} \propto \frac{\text{surface area} \times \text{difference in concentration}}{\text{thickness of exchange surface}}$$

Alveoli have **adaptations** for each of the **three factors** in Fick's law, which make them a really good surface for gas exchange (see p.22):

- The huge number of alveoli means there's a **big surface area** for exchange.
- They have a **thin lining** (the **alveolar epithelium**) so there's a **short diffusion pathway**.
- They also **maintain steep concentration gradients** of the respiratory gases to speed up diffusion:

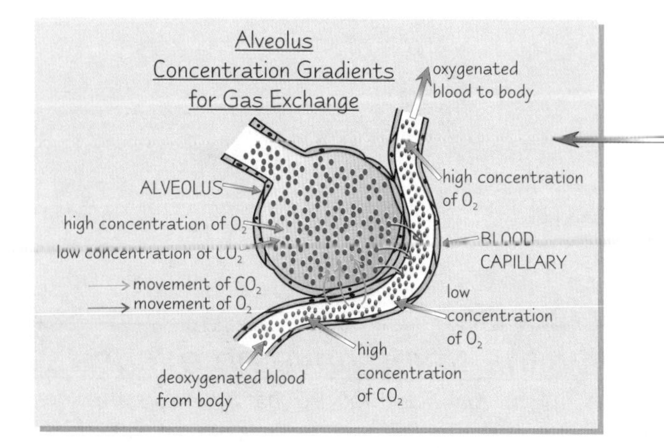

After exchanging gases with cells, **deoxygenated blood** returns to blood capillaries in the lungs. There's a higher level of **O_2** in the **alveoli** compared to the capillaries. This gives a good concentration gradient for **diffusion of oxygen** into the **blood**. The low level of CO_2 in the **alveolar space** helps to remove the CO_2 from the blood. **CO_2 diffuses** down the concentration gradient and **into the lungs**, where it's breathed out.

Lungs and Ventilation

Ventilation is the Act of Breathing In and Breathing Out

We need to get gases inside our bodies from the external environment, and transfer waste gases out of our bodies. We need **ventilation** to do this, which consists of **inspiration** (breathing in) and **expiration** (breathing out).

1) **Inspiration** is an **active process**. The **intercostal and diaphragm muscles contract**. This causes the **rib cage to move upwards and outwards** and the **diaphragm to flatten**. This **increases the volume** in the thorax (the space where the lungs are), which **decreases air pressure** to below atmospheric pressure, so **air flows into the lungs**.
2) **Expiration** is mainly a **passive process**. The **intercostal and diaphragm muscles relax,** so the **rib cage moves downwards and inwards** and the **diaphragm becomes curved again**. This **decreases the volume** in the thorax, which **increases air pressure** to above atmospheric pressure, so **air is forced out of the lungs**.

Inspiration

air flows in

volume increases therefore air pressure decreases

intercostal muscles contract, causing ribs to move outwards and upwards

diaphragm muscles contract, causing diaphragm to move downwards and flatten

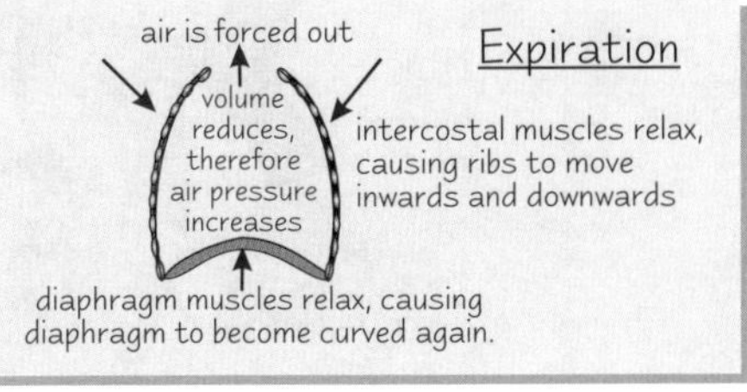

The table shows the difference in composition between the air you breathe in and the air you breathe out again a moment later — the main difference is you've used up some oxygen and added a bit of CO_2.

Gas	% Inhaled	% Exhaled
Oxygen	20.96	16.50
Carbon Dioxide	0.04	4.00
Nitrogen	79.00	79.00
Water Vapour	Variable	Saturated

The Medulla Oblongata in the Brain Controls the Rate of Breathing

Your **breathing rate** changes according to how much **physical activity** you're doing. When you're exercising you use more **energy**. Your body needs to do more **aerobic respiration** to release this energy, so it needs more **oxygen**.

The **ventilation cycle** is the cycle of breathing in and out. It involves **inspiratory** and **expiratory** centres in the **medulla oblongata** (an area of the brain) and **stretch receptors** in the lungs.

The Ventilation Cycle:

1) The medulla's **inspiratory centre** sends nerve impulses (via **phrenic nerves**) to the **intercostal** (rib) and **diaphragm** muscles to make them **contract**.
   It also sends nerve impulses to the medulla to **inhibit** the **expiratory centre**.
2) The **lungs inflate**. This stimulates **stretch receptors**, which send nerve impulses back to the **medulla** to **inhibit** the **inspiratory centre**.
3) Now the expiratory centre (no longer inhibited) sends nerve impulses to the muscles to relax and the **lungs deflate**, expelling air. This causes the **stretch receptors** to become **inactive**, so the inspiratory centre is no longer inhibited and the cycle starts again.

Practice Questions

Q1 What does the trachea branch into in the lungs?

Q2 Name three ways that the alveoli are adapted for gas exchange.

Q3 Explain what happens during inspiration.

Q4 Which bit of the brain controls breathing rate?

Exam Question

Q1 Describe how the alveoli in the lungs are adapted for gas exchange, with reference to Fick's law. [4 marks]

Alveoli — useful things...always make me think of pasta...

I know you've just got to the end of a page — but it would be a pretty smart idea to have another look at Transport Across Membranes on page 7. It's all the stuff about Fick's Law, and transport across membranes. Not the most thrilling prospect I realise, but it'll help the stuff on gas exchange in the alveoli make more sense.

The Heart and the Cardiac Cycle

*The heart keeps blood pumping **continually** round your body. It's happening right now as you read this rather long and dull sentence that I'm writing to keep you reading without stopping to highlight the **unceasingness** of blood flow.*

*The **Heart** Consists of **Two Muscular Pumps***

The diagram below shows the **internal structure** of the heart. The **right side** pumps **deoxygenated blood** to the **lungs** and the **left side** pumps **oxygenated blood** to the **whole body**. NB — the **left and right side** are **reversed** on the diagram, 'cos it's the left and right of the person that the heart belongs to.

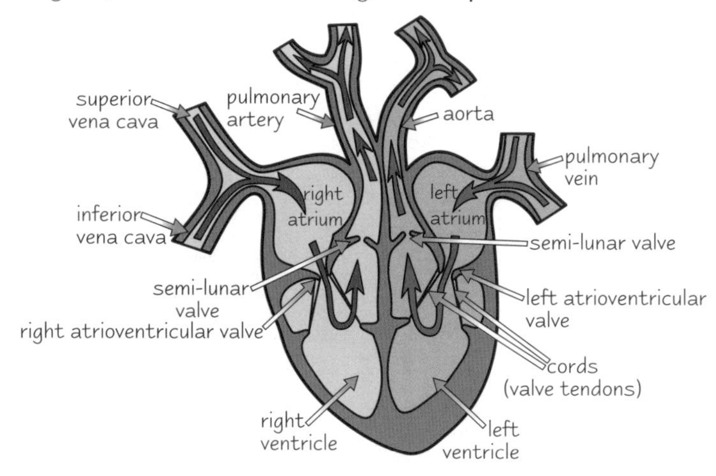

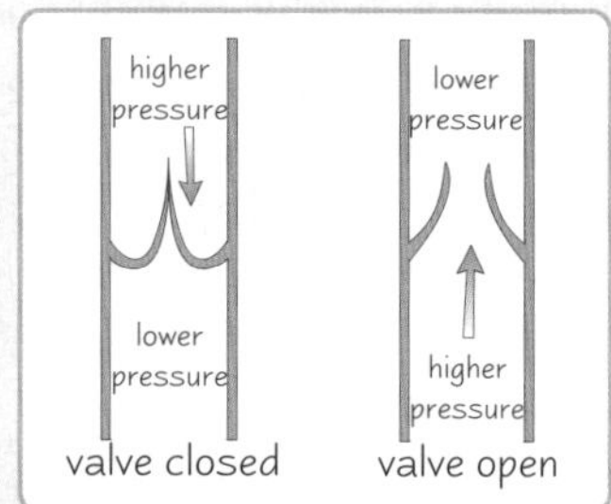

The **valves** only open one way — whether they open or close depends on the relative pressure of the heart chambers. If there's higher pressure behind a valve, it's forced open, but if pressure is higher above the valve it's forced shut.

Each bit of the heart is adapted to do its job effectively.

1) The **left ventricle** of the heart has thicker, more muscular walls than the **right ventricle**, because it needs to contract powerfully to pump blood all the way round the body. The right side only needs to get blood to the lungs, which are nearby.
2) The **ventricles** have thicker walls than the **atria**, because they have to push blood out of the heart whereas the atria just need to push blood a short distance into the ventricles.
3) The **atrioventricular valves** link the atria to the ventricles and stop blood getting back into the atria when the ventricles contract.
4) The **semilunar valves** stop blood flowing back into the heart after the ventricles contract.
5) The **cords** attach the atrioventricular valves to the ventricles to stop them being forced up into the atria when the ventricles contract.

***Cardiac Muscle** Controls the **Regular Beating** of the Heart*

Cardiac (heart) muscle is '**myogenic**' — this means that, rather than receiving signals from **nerves**, it contracts and relaxes on its own. This pattern of contractions controls the **regular heartbeat**.

1) The process starts in the **sino-atrial node (SAN)** in the wall of the **right atrium**.
2) The SAN is like a pacemaker — it sets the rhythm of the heart beat by sending out regular **electrical impulses** to the atrial walls.
3) This causes the right and left atria to contract **at the same time**.
4) A band of non-conducting **collagen tissue** prevents the electrical impulses from passing directly from the atria to the ventricles.
5) Instead, the **atrioventricular node (AVN)** picks up the impulses from the SAN. There is a **slight delay** before it reacts, so that the ventricles contract **after** the atria.
6) The AVN generates its own **electrical impulse**. This travels through a group of fibres called the **bundle of His** and then into the finer fibrous tissue in the right and left ventricle walls.
7) The impulses mean both ventricles **contract simultaneously**, from the bottom up.

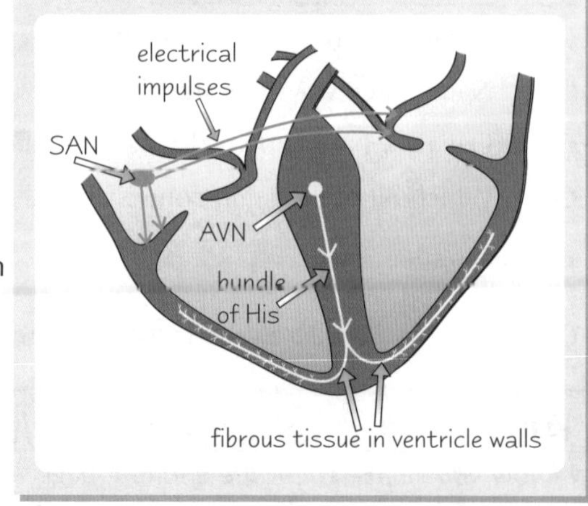

The Heart and the Cardiac Cycle

The **Cardiac Cycle** Pumps Blood Round the Body

The cardiac cycle is an ongoing sequence of **contraction** and **relaxation** of the atria and ventricles that keeps blood continuously circulating round the body. The contraction and relaxation patterns alter the **volume** of the different heart chambers, which alters **pressure** inside the chambers. This causes **valves** to open and close, which directs the **blood flow** through the system. There are 3 stages:

1) Ventricles relax, atria contract

The **ventricles both relax**. The atria then contract, which decreases their volume. The resultant higher pressure in the atria causes the atrioventricular valves to open. This forces blood through the valves into the ventricles.

2) Ventricles contract, atria relax

The **atria relax** and the **ventricles then contract**. This means pressure is higher in the ventricles than the atria, which shuts off the atrioventricular valves to prevent back-flow. Meanwhile, the high pressure opens the semilunar valves and blood is forced out into the pulmonary artery and aorta.

3) Ventricles relax, atria relax

The **ventricles and the atria both relax**, which increases volume and lowers pressure in the heart chambers. The higher pressure in the pulmonary artery and aorta closes the semilunar valves to prevent back-flow. Then the atria fill with blood again due to higher pressure in the vena cava and pulmonary vein and the cycle starts over again.

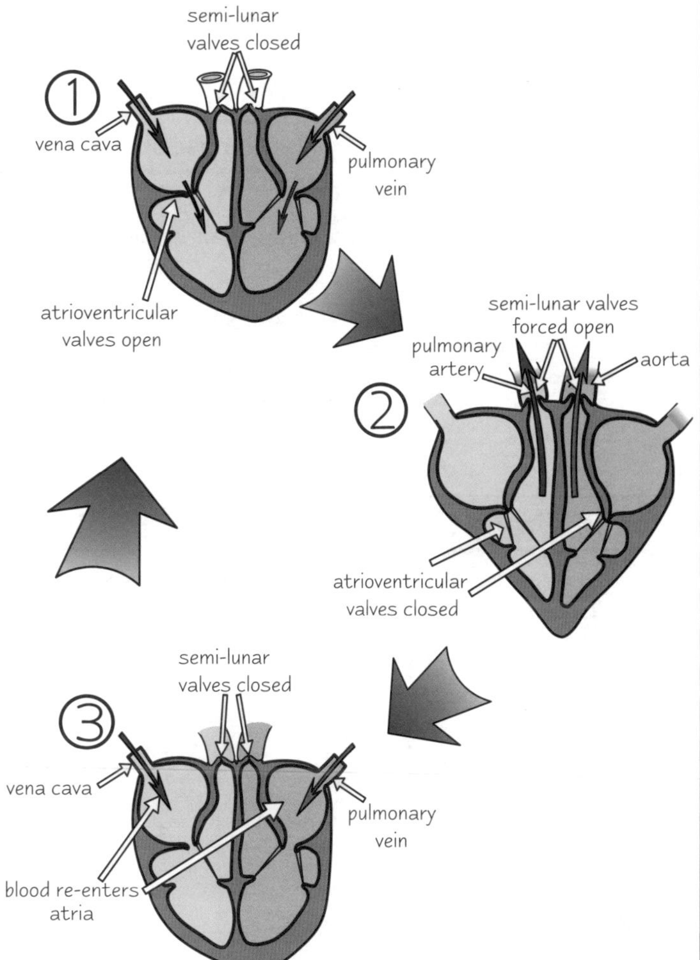

Practice Questions

Q1 Which side of the heart carries oxygenated blood?

Q2 Why is the left ventricle more muscular than the right ventricle?

Q3 What is the purpose of heart valves?

Q4 What does "myogenic" mean?

Q5 What is the difference between the SAN and the AVN?

Exam Questions

Q1 Describe the pressure changes which occur in the heart during contraction and relaxation. [3 marks]

Q2 Explain how valves stop blood going back the wrong way. [6 marks]

Learn these pages off by heart…

Some of this will be familiar to you from GCSEs, but the stuff on controlling heartbeat is all shiny and new. Just think, you don't have to think consciously about making your heart beat — your body does it for you. So you couldn't stop it beating even if for some strange reason you wanted to. Which is nice to know.

Effects of Exercise

A person's breathing rate and heart rate change when they exercise. For some reason, biologists see this as an excuse to start using technical terms and sticking them in equations. Don't let that bother you — it's all pretty straightforward.

Learn the Equations for Cardiac Output and Pulmonary Ventilation

1) **Heart Rate** — the number of heartbeats per minute. You can measure your heart rate by feeling your pulse, which is basically surges of blood forced through the arteries by the heart contracting. During exercise the heart rate speeds up. Afterwards it returns to resting rate (around 70 beats per minute).
2) **Stroke Volume** — the volume of blood pumped during each heartbeat, measured in cm^3.
3) **Cardiac Output** — the amount of blood pumped by the heart per minute (measured in cm^3 per minute).

cardiac output = stroke volume × heart rate

1) **Tidal Volume** is the volume of air in each breath (measured in cm^3).
2) **Breathing Rate** is the number of breaths per minute. For a person at rest it's about 15 breaths.
3) **Pulmonary ventilation** is the volume of air taken into the lungs in one minute. It's measured in cm^3 per min.

pulmonary ventilation = tidal volume × breathing rate

The Brain Controls Changes in the Heart Rate due to Exercise

When we exercise, our **cardiac output increases**. Nerve impulses from the brain **modify the heart rate** — for example by making the heart beat faster if we are exercising.

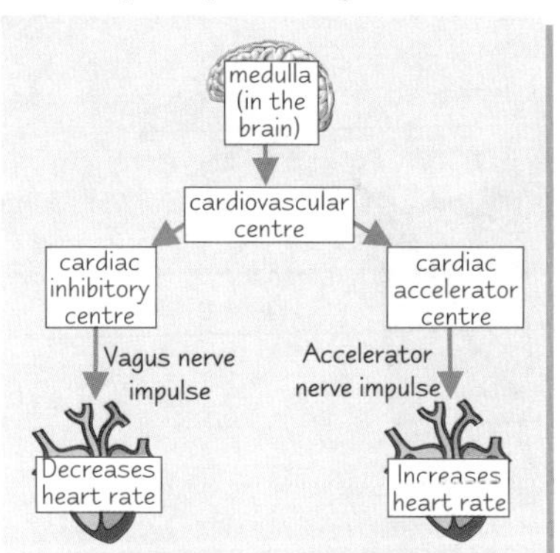

1) **Heart rate** is controlled in the **cardiovascular centre** in the **medulla oblongata** area of the **brain**.
2) **Nerve impulses** from the **cardiovascular centre** reach the **SAN**.
3) An **accelerator nerve impulse** from the cardiac accelerator centre stimulates the SAN to **increase the heartbeat**.
4) A **vagus nerve impulse** from the cardiac inhibitory centre stimulates the SAN to **slow the heart rate down**.

Doug's medulla oblongata was having trouble controlling his heart rate.

During exercise, blood flow is redistributed round the body:

1) Blood flow to the **skeletal muscles** and **heart increases** (don't forget that the heart is a muscle too, so it needs a constant supply of blood to function properly and pump blood to other areas).
2) The blood flow to the **digestive system decreases** (so more is available for skeletal muscles).
3) The blood flow to the **brain** and **kidneys stays the same**. This is because they both need a **constant supply** of blood to function. If the **kidneys** didn't receive enough blood, **toxins** in the blood would build up. If the **brain** didn't receive enough, we could become **unconscious** or even **die**.

Pressure Receptors Inform the Brain About Changing Blood Pressure

Pressure receptors inform the **brain** about changes in **blood pressure** so it can **alter heart rate** accordingly.

1) It's important that **blood pressure** is maintained at a constant level. **Pressure receptors** are found in the **aorta wall** and in **carotid arteries**. They detect changes in arterial blood pressure and inform the brain.
2) If the pressure is **too high**, they send impulses to the cardiovascular centre. In turn, this sends impulses to the SAN, to **slow down heart rate**.
3) If **pressure is too low**, pressure receptors send impulses to the cardiovascular centre, which sends its own impulses to the SAN, to **speed up heart rate**.

Effects of Exercise

Chemoreceptors Inform the Brain about Changing Blood pH

Chemoreceptors inform the brain about changes in **blood pH** so it can (predictably) **alter heart rate** accordingly.

1) An **increase in respiration**, for example due to physical activity, increases the **level of CO_2** in the blood. Increased CO_2 levels **lower the pH of blood**, which is detected by **chemoreceptors** in the **carotid artery walls**.
2) In the same way as pressure receptors, these send impulses to the brain, which sends its own impulses to the SAN to **increase the heart rate**.

This diagram shows how both pressure receptors (see p.30) and chemoreceptors affect heart rate by sending messages to the cardiovascular centre.

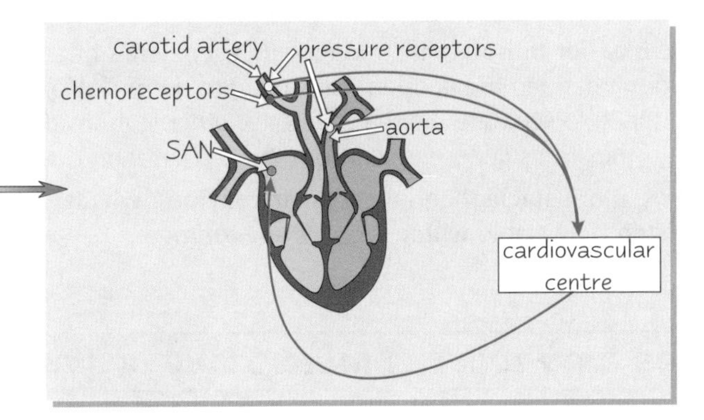

The Brain also Controls Changes in Pulmonary Ventilation due to Exercise

When we **exercise**, our **pulmonary ventilation increases** as well as our cardiac output.
This is because we breathe faster and more deeply, to get oxygen to our respiring muscles more quickly.
Chemoreceptors detect a **decrease in blood pH** when we exercise and inform the brain about it.

1) The **chemoreceptors** are found in the **medulla oblongata**, in **aortic bodies** (in the aorta), and in **carotid bodies** (in the carotid arteries carrying blood to the head).
2) If these chemoreceptors **detect** a **decrease** in the **pH** of the blood, they send a **signal** to the **medulla** to send more frequent nerve impulses to the **intercostal muscles** and **diaphragm**. This **increases** the **rate** of **breathing** and the **depth** of breathing.
3) This allows **gaseous exchange** to **speed up**. CO_2 levels drop and the demand for extra O_2 by the muscles is met.

Practice Questions

Q1 How would you calculate cardiac output?

Q2 What is tidal volume?

Q3 Which part of the brain modifies heart rate?

Q4 How do pressure receptors help control heart rate?

Q5 What structures detect changes in pH in the blood?

Exam Questions

Q1 Explain how and why distribution of blood flow changes during exercise. [4 marks]

Q2 Describe what would happen to the heart rate if the heart suffered a sudden loss of blood pressure. [4 marks]

Q3 How would you expect pulmonary ventilation to change when an athlete starts to run a race? Explain your answer. [5 marks]

Tidal volume — nothing to do with the sound of the sea...

I may have been a bit optimistic when I said this was straightforward. OK, I was lying. There are loads of terms to learn, and it's tempting just to read them and hope for the best. But the safest way is to close the book and try to write them out, along with their meanings. Keep doing this until the cows come home and the fat lady sings — or, until you know them all.

Isolation of Enzymes

Enzymes' special properties mean that they are used loads in biotechnology (biotechnology is the use of micro-organisms to make useful products). Because enzymes are just so darn important, many of them are mass-produced industrially.

Extracellular *enzymes are used in* ***Biotechnology***

Enzymes that work **inside cells** are **intracellular enzymes**. Ones that **pass out** of cells are **extracellular enzymes** — these are used more often in **biotechnology** for two main reasons:

1) They're **easier to isolate** than intracellular enzymes because you don't need to **break open** cells to get to them. Also, you don't need to **separate** them from all the other stuff inside a cell — they're usually secreted from the cell on their own.
2) They're **more stable** than intracellular enzymes, which are often only stable within the cell environment.

Many **micro-organisms** (e.g. some bacteria and fungi) secrete **extracellular enzymes** onto their food to digest it. Because of this, bacteria are often used to **commercially produce** large quantities of **useful extracellular enzymes** that can then be used to create useful products.

E.g. *Bacillus subtilis* bacteria are used for the commercial production of extracellular protease enzymes.

Enzymes *are Produced* ***Industrially*** *by* ***Fermentation***

In biotechnology, **fermentation** is the process of **culturing** (growing) micro-organisms to produce **enzymes**, like proteases, pectinases and lactases.

1) The bacteria (e.g. *Bacillus subtilis*) are grown in large, non-corrosive vessels called **fermenters**.
2) Inside the fermenter the nutrient medium, temperature, pH and oxygen levels are carefully **monitored**, to provide **optimum** conditions for the bacteria's growth. The faster the bacteria grow, the more enzyme will be produced.
3) The bacteria are grown under **aseptic conditions** to avoid contamination by other micro-organisms. This involves:
   - checking the **starter culture** of bacteria is pure — removing any unwanted micro-organisms;
   - sterilising the **fermenter** with steam before use;
   - sterilising the **nutrient medium** in the fermenter before the bacteria are added;
   - sterilising the **air** supplied to the fermenter.

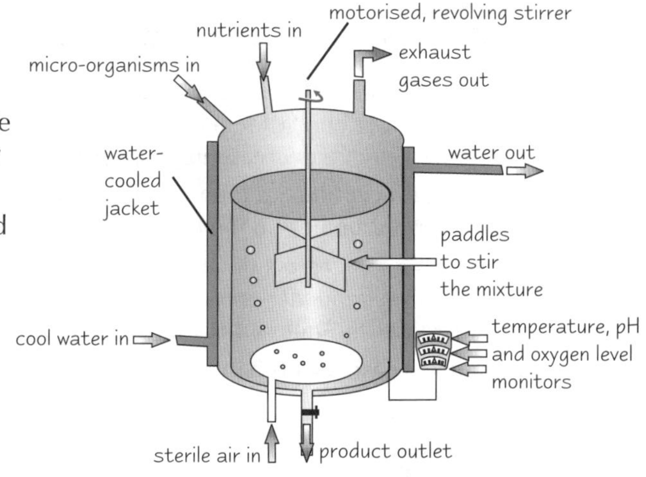

4) A process called **downstream processing** is then used to **extract** and **purify** of the enzyme, turning it from bacteria-and-enzyme slush into pure, dry, powdered enzyme.

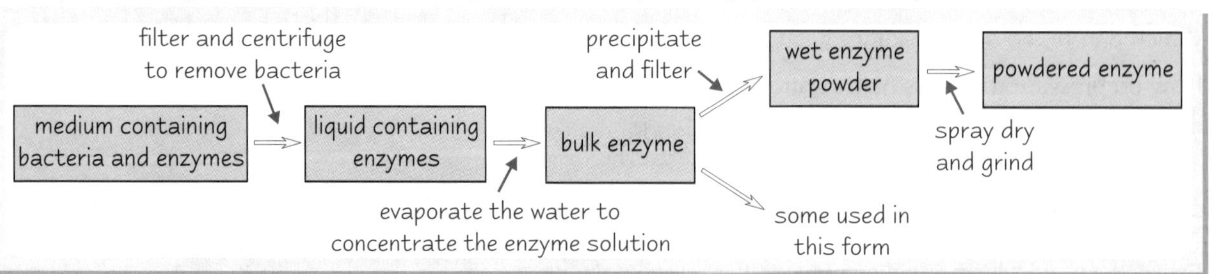

Proteases *are Found in* ***Washing Powders***

Biological washing powders and liquids contain enzymes to help them **remove stains**.

1) **Proteases** digest (**hydrolyse**) proteins in milk, egg and blood.
2) **Lipases** digest fats.
3) **Cellulases** are sometimes added to get rid of the fluff that dulls the colour of worn fabrics.

Enzymes added to detergents must work well at pH 9-11, and must be thermostable (stable in heat) so they'll work in hot water.

Isolation of Enzymes

Thermostable Enzymes can Tolerate High Temperatures

Some enzymes found in micro-organisms living in **very hot environments** (e.g. **volcanoes**) can tolerate much higher temperatures than other enzymes. These are **thermostable enzymes** — they're **useful** in **industrial processes** because:

- **Reactions** happen **faster** at higher temperatures, so the product is produced more quickly.
- The temperature in the fermenter doesn't have to be so carefully **monitored**, which **saves money**.

Enzymes can be used to Detect Specific Chemicals

Enzymes' properties mean that many are used as **medical analytical reagents**.

Enzymes are useful because:

1) They're **specific** (they only bind to one test substance), so they're useful for **identifying specific substances**.
2) They're very **sensitive** and react with **low concentrations** of substrate.
3) They **work quickly**.

A simple test for **diabetes mellitus** uses a **reagent strip** which detects glucose in urine. The plastic strip has a coloured square at one end. The square contains **3 reagents**:

- **glucose oxidase**, an enzyme which catalyses the reaction

  glucose + oxygen + water ⟹ gluconic acid + hydrogen peroxide

- **pink dye** which changes colour when it's oxidised by hydrogen peroxide
- **peroxidase**, an enzyme which catalyses the oxidation of the dye

  hydrogen peroxide + pink dye ⟹ blue dye + water

The strip is dipped into a sample of urine — if it turns **blue**, glucose is present.

Immobilised Enzymes have many Advantages

Immobilised enzymes are attached to or **trapped** in a non-reactive, insoluble material, like a **fibrous polymer mesh.**

1) The enzymes aren't mixed in with the product so it's easy to recover them and **re-use** them. This keeps **production costs** down.
2) The product isn't **contaminated** by the enzyme.
3) Immobilised enzymes are more **pH-** and **heat-stable**. E.g. The **dairy industry** uses **immobilised lactase** to convert the lactose in milk into galactose and glucose.

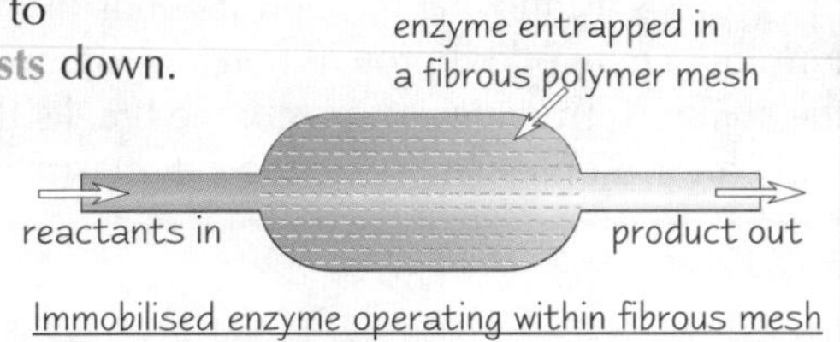

Immobilised enzyme operating within fibrous mesh

Practice Questions

Q1 What is the difference between intracellular and extracellular enzymes?

Q2 Name three conditions inside a fermenter that are monitored and controlled during the fermentation process.

Q3 What do manufacturers add to the detergent in biological washing powders?

Q4 What is a thermostable enzyme?

Exam Questions

Q1 a) Explain why extracellular rather than intracellular enzymes are usually used in commercial processes. [3 marks]

b) Explain why it is important to prevent microbial contamination during fermentation. [2 marks]

Q2 a) What is an immobilised enzyme? [2 marks]

b) Explain two advantages of using immobilised enzymes. [2 marks]

Where do you put a horse to keep it warm? — In a thermostable...

Enzymes are grrreat — if they didn't exist we'd all have smeary stains down the fronts of our clothes. OK, so we'd also all be dead, but you have to get your priorities right — dirty clothes won't get you far in life. One of the greatest lessons my mum taught me was that it's hard to make new friends, find romance and get a job if you've got food all down your clothes.

Mitosis and the Cell Cycle

I don't like cell division. There, I've said it. It's unfair of me, because if it wasn't for cell division I'd still only be one cell big. It's all those diagrams that look like worms nailed to bits of string that put me off.

Mitosis is **Cell Division** that Produces **Genetically Identical Cells**

Cells increase in number by **cell division**. There are two types of cell division — **mitosis** and **meiosis**. **Mitosis** produces daughter cells that are **genetically identical** to the parent cell. It's used in **asexual reproduction**. It's also needed for the **growth** of multicellular organisms (like us) and for **repairing** damaged tissues. How else do you think we get from being a baby to being a big, strapping lass / lad — it's because the cells in our bodies divide and multiply.

Mitosis has **Four Division Stages** plus **Interphase**

Mitosis is really one **continuous process**, but it's described as a series of **division stages** — prophase, metaphase, anaphase and telophase. **Interphase** comes before the division stages — it's when cells grow and prepare to divide by replicating their DNA.

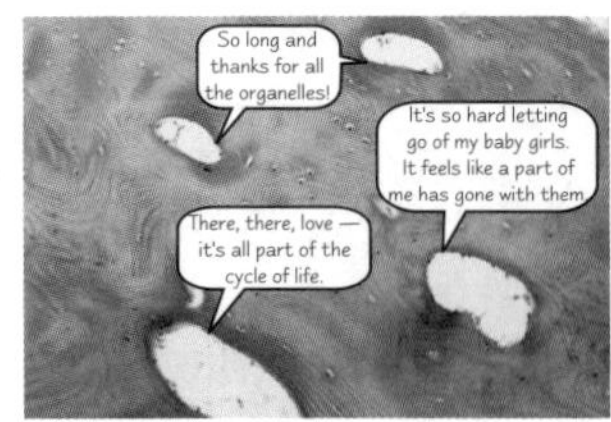

Mitosis can be a moving time.

1) **Interphase** — The cell carries out normal functions, but also prepares to divide. It replicates its DNA, to double its genetic content. It also replicates its organelles so it has spare ones, and increases its ATP content (ATP provides the energy needed for cell division).

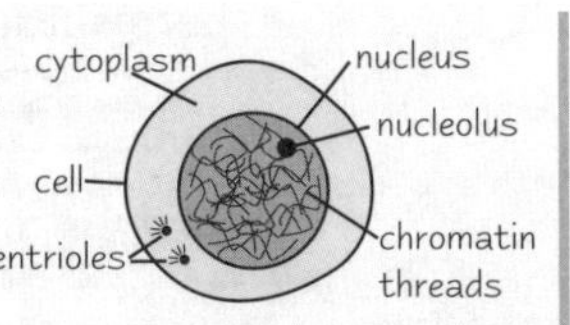

2) **Prophase** — The chromosomes **condense**, getting shorter and fatter. Tiny bundles of protein called **centrioles** start moving to opposite ends of the cell, forming a network of protein fibres across it called the **spindle**. The nuclear membrane breaks down and chromosomes lie free in the cytoplasm.

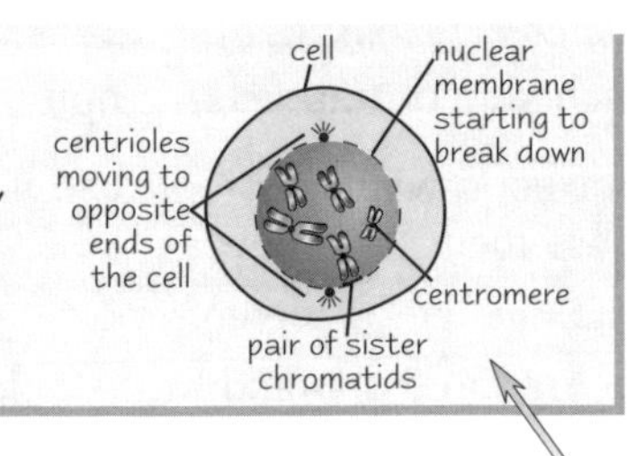

As mitosis begins, the chromosomes are made of two strands joined in the middle by a centromere. The separate strands are called chromatids. There are two strands because each chromosome has already made an identical copy of itself during interphase. When mitosis is over, the chromatids end up as one-strand chromosomes in the new daughter cells.

3) **Metaphase** — The chromosomes (each with two chromatids) line up along the middle of the cell and become attached to the spindle by their centromere.

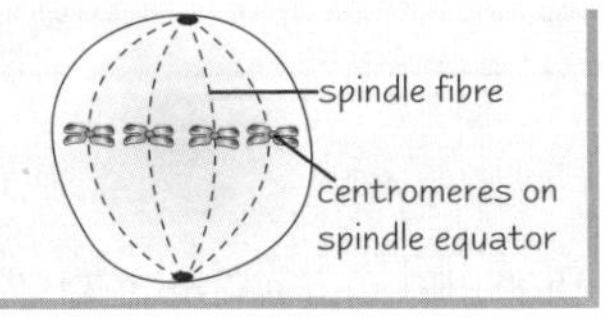

4) **Anaphase** — The centromeres attaching the chromatids to the spindles divide, separating each pair of sister chromatids. The spindles contract, pulling chromatids to opposite poles, centromere first.

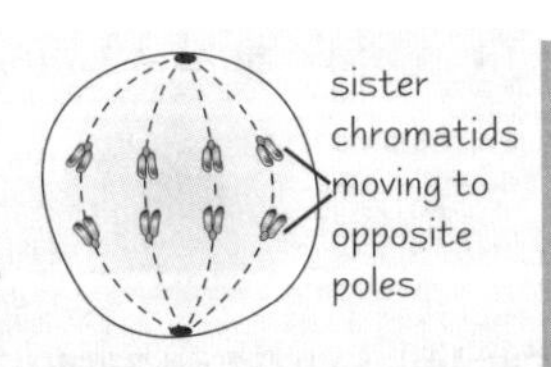

5) **Telophase** — The chromatids reach the **opposite poles** on the spindle. They uncoil and become long and thin again. They're now called **chromosomes** again. A **nuclear membrane** forms around each group of chromosomes, so there are now **two nuclei**. The **cytoplasm divides** and there are now **two daughter cells** which are **identical** to the original cell. Mitosis is finished and each daughter cell starts the **interphase** part of the cell cycle to get ready for the next round of mitosis.

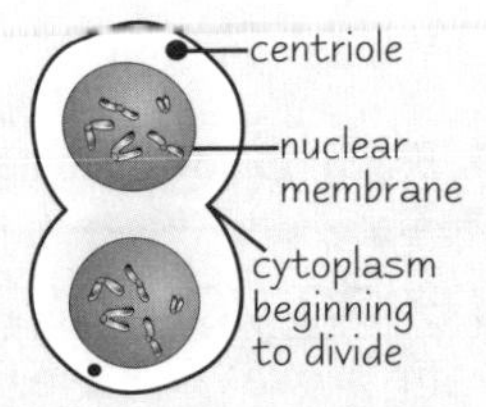

Mitosis and the Cell Cycle

All Cells in Multi-cellular Organisms Follow the Same Cycle

Cells from multi-cellular organisms have a clear **cell cycle** that starts when they are produced by cell division, and ends with them dividing themselves to produce more identical cells. The cell cycle consists of a period of cell division (**mitosis**), and a period in between divisions called **interphase**. Interphase is sub-divided into 3 separate growth stages. These are called $\mathbf{G_1}$, **S** and $\mathbf{G_2}$. Each stage involves specific cell activities:

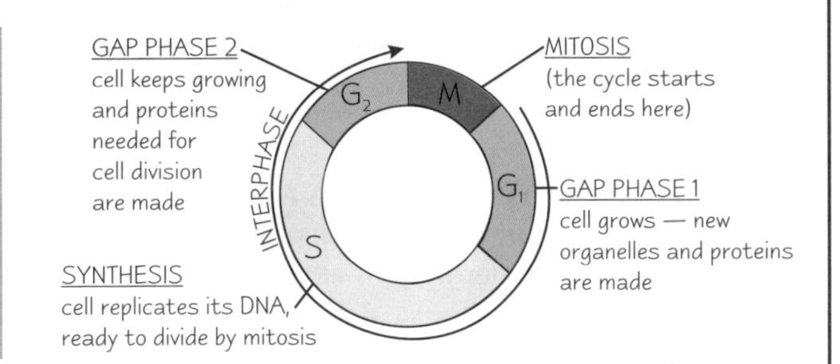

You can Observe Mitosis by Staining Chromosomes

You can **stain chromosomes** so you can see them under a **microscope**. This means you can watch what happens to them **during mitosis** — and it makes high-adrenaline viewing, I can tell you.

For example, to see the chromosomes in **onion cells**, you use the following technique:

1) Put a small piece of onion in a half-and-half mixture of **ethanoic acid** and **alcohol**, and leave for 10 minutes.
2) Then rinse it with water and put it in a **hydrochloric acid** solution. Warm it in here for 5 minutes, to separate all the cells from each other.
3) Then rinse off the onion sample with water again and put it on a slide with a dye called **ethanoic orcein**. This is what stains the chromosomes.
4) Then put a coverslip on top of the sample and gently squash it down to squish the cells into a **thin layer**. This makes the cells and chromosomes easy to see.

Practice Questions

Q1 What is cell division by mitosis needed for?

Q2 Explain why the chromosome becomes visible as two chromatids at the end of prophase.

Q3 What happens during the "synthesis" stage of interphase in the cell cycle?

Exam Questions

Q1 The diagrams show cells at different stages of mitosis.

Cell A

Z X Y

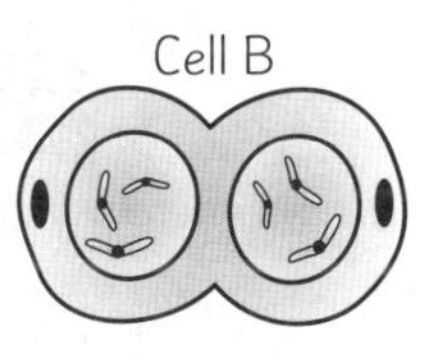

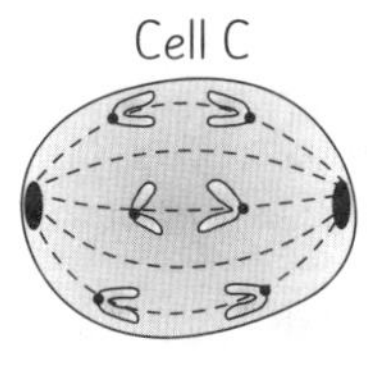

a) For each of the cells A, B and C state the stage of mitosis, giving a reason for your answer. [6 marks]

b) Name the structures labelled X, Y and Z in cell A. [3 marks]

Q2 During which stages of the cell cycle would the following events take place?

a) DNA replication. [2 marks]

b) Formation of spindle fibres. [2 marks]

Doctor, I'm getting short and fat — don't worry, it's just a phase...

Quite a lot to learn in this topic — but it's all dead important stuff so no slacking. All cells undergo mitosis — it's how they multiply and how organisms like us grow and develop. Remember that chromosomes are in fact usually made up of two sister chromatids joined by a centromere. Aaw, nice to know family values are important to genetic material too.

Meiosis

More cell division — lovely jubbly. Meiosis is the cell division used in sexual reproduction. It consists of two divisions, not one.

The **Haploid Number** of **Chromosomes** is **Half** the **Diploid Number**

A chromosome is a thread-like structure made up of **one long, tightly-coiled molecule** of **DNA**.

1) Each human cell contains 46 single chromosomes in total — this is called the **diploid number** (**2n**). Different species have different diploid numbers. Chromosomes in normal body cells can be grouped into **homologous pairs** (matching pairs).
2) **Gametes** (sex cells) contain **half** the diploid number of chromosomes (i.e. one chromosome from each homologous pair) — this is called the **haploid number** (**n**). Why this happens is explained at the bottom of the page.

More on chromosomes:

The **DNA molecule** in a chromosome is divided into lots of sections. Each section carries the code to make a particular **protein**. These sections are called **genes**. In a **homologous pair** of chromosomes, both chromosomes are the same shape and size and have the **same genes** in the same locations. One chromosome in each homologous pair comes from the **female parent**, and the other from the **male parent**.

DNA From **One Generation** is **Passed** to the Next by **Gametes**

1) **Gametes** are the **reproductive cells** (**sperm** in males and **ova** in females). They join together at **fertilisation** to form a **zygote**, which divides and develops into a **new organism**.
2) **Gametes** have a **haploid** (**n**) number of chromosomes — there's only **one version** of each chromosome.
3) At **fertilisation**, a **haploid sperm** fuses with a **haploid ovum**, making a cell with the normal diploid number of chromosomes. So half the chromosomes in this diploid cell have come from the father (the sperm) and half from the mother (the ovum).

Meiosis Halves the **Chromosome Number**

Meiosis is a type of cell division. It's essential for **sexual reproduction**. During meiosis, the diploid number of chromosomes in a cell is reduced to the haploid number in the daughter cells:

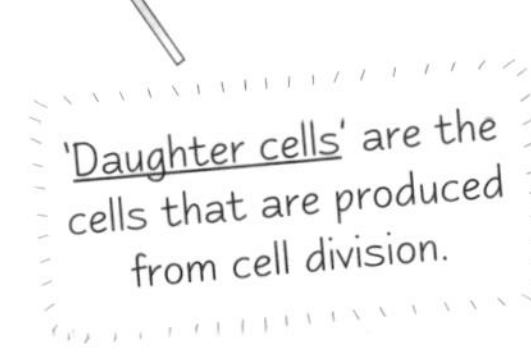

1) Cells that divide by meiosis are **diploid** to start with, but the daughter cells are **haploid**. This makes sure cells possess a **constant number of chromosomes** through the generations — without meiosis, you'd get **double** the number of chromosomes in each generation, when the gametes fused.
2) Meiosis happens in the **reproductive organs**. In humans it's in the **testes** for males and the **ovaries** for females. In plants it's in the **anthers** and **ovules**.
3) Unlike mitosis, there are **two divisions**. This **halves** the chromosome number.

Meiosis

Meiosis I *Separates Chromosomes,* ***Meiosis II*** *Separates Chromatids*

There are **two divisions** in meiosis. These divisions are called **meiosis I** and **meiosis II**.

1) In **Meiosis I** the **homologous pairs** of **chromosomes** are separated, which **halves** the number of chromosomes in the daughter cells.
2) **Meiosis II** is like mitosis — it separates the **pairs of chromatids** that make up each chromosome.
3) Unlike mitosis, which results in two genetically identical diploid cells, meiosis results in **four haploid cells** (gametes) that **are genetically different** from each other. So meiosis is important for producing **genetic variation** in organisms.

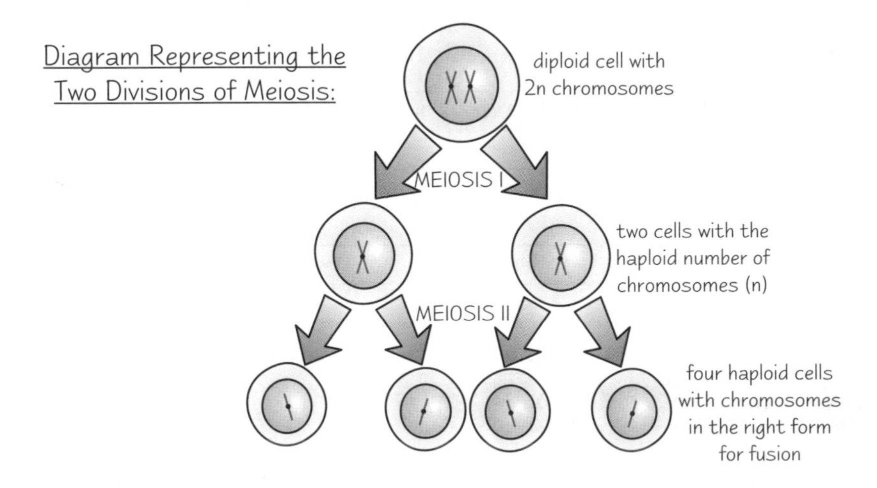

Practice Questions

Q1 Explain what is meant by the terms 'haploid' and 'diploid'.

Q2 In which cells in a human would meiosis take place?

Q3 How many divisions are there in meiosis?

Q4 What happens during meiosis I?

Exam Question

Q1 The diagram shows stages of meiosis in a human ovary. Each circle represents a cell.

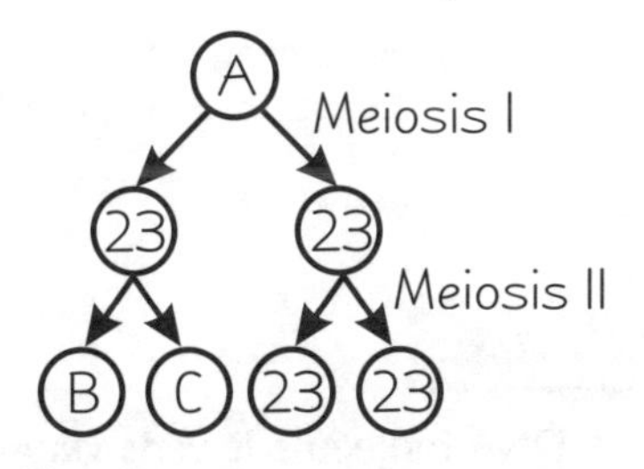

a) How many chromosomes would be found in cells A, B and C? [3 marks]

b) Explain why it's important for gametes to have half the number of chromosomes as normal body cells. [2 marks]

Reproduction isn't as exciting as some people would have you believe...

For some reason, this stuff can take a while to go in (insert own joke). But that's no excuse to just sit there staring frantically at the page and muttering "I don't get it," over and over again. Use the diagram to help you understand — it might look evil, but it really helps. The key thing is to understand what happens to the ***number of chromosomes*** *in meiosis.*

Basic Structure of DNA and RNA

*These pages are about the structure of DNA (**deoxy**ribonucleic acid) and RNA (plain ol' ribonucleic acid), plus a little thing called DNA self replication, which is kinda important to us living things. (OK, spot the major understatement here.)*

DNA and RNA are Very Similar Molecules

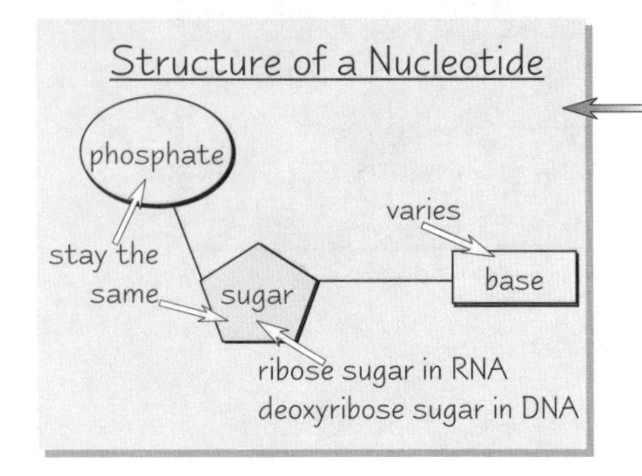

DNA and RNA are **nucleic acids** — made up of lots of **nucleotides** joined together. Nucleotides are units made from a **pentose sugar** (with 5 carbon atoms), a **phosphate** group and a **base** (containing nitrogen and carbon).

The sugar in **DNA** nucleotides is a **deoxyribose** sugar — in **RNA** nucleotides it's a **ribose** sugar. Within DNA and RNA, the sugar and the phosphate are the same for all the nucleotides. The only bit that's different between them is the **base**. There are five possible bases found in DNA and RNA:

Although DNA is called deoxyribonucleic acid, it still contains oxygen.

BASE	found in DNA	found in RNA
adenine	✓	✓
guanine	✓	✓
cytosine	✓	✓
thymine	✓	×
uracil	×	✓

DNA and RNA are Polymers of Mononucleotides

Mononucleotides (single nucleotides) join together by a **condensation reaction** between the **phosphate** of one group and the **sugar** molecule of another. As in all condensation reactions, **water** is a by-product.

DNA is made of **two strands of nucleotides**. **RNA** has just the one strand.
In DNA, the strands spiral together to form a **double helix**.
The strands are held together by **hydrogen bonds** between the bases.

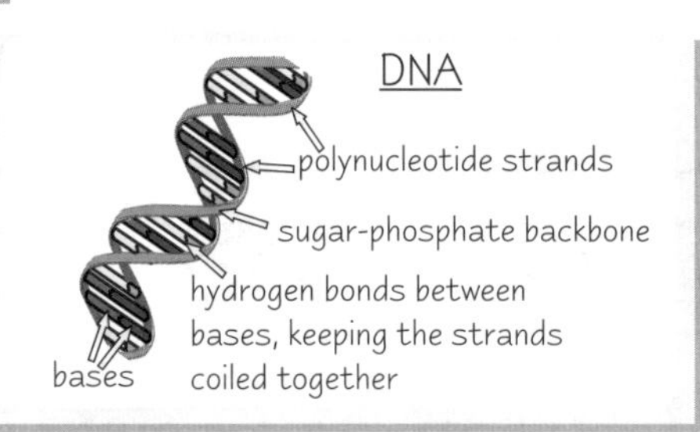

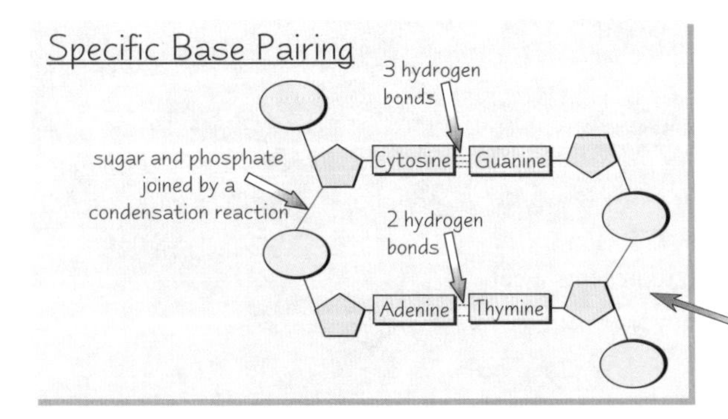

Each base can only join with one particular partner — this is called **specific base pairing**.

1) In DNA **adenine** always pairs with **thymine** (**A - T**) and **guanine** always pairs with **cytosine** (**G - C**).
2) It's the same in RNA, but **thymine**'s replaced by **uracil** (so it's **A - U** and **G - C**).

2 hydrogen bonds form between adenine and thymine / uracil).
3 hydrogen bonds form between guanine and cytosine.

DNA's Structure Makes it Good at its Job

1) The job of DNA is to carry **genetic information**. A DNA molecule is very, very **long** and is **coiled** up very tightly, so a lot of genetic information can fit into a **small space** in the cell nucleus.
2) Its **paired structure** means it can **copy itself** — this is called **self-replication** (see p.39).
It's important for cell division and for passing on genetic information to the next generation.

In the 1950s, scientists proved that DNA is involved in transmitting genetic information by studying how viruses replicate inside bacterial cells. By labelling the viral DNA with a radioactive substance that could be tracked, the scientists observed that the viruses injected their DNA into cells. The DNA then replicated inside the cells. The new viruses were genetically identical to the original virus, showing that genetic information had been passed on.

Replication of DNA

DNA can Copy Itself — Self-Replication

DNA has to be able to **copy itself** before **cell division** can take place, which is essential for growth and development and reproduction — pretty important stuff.

1) **Specific base pairing** means that each type of base in DNA only pairs up with one other type of base — **A** with **T**, **C** with **G**. When a molecule of **DNA splits**, the **unpaired bases** on each strand can match up with complementary bases on **free-floating nucleotides** in the cytoplasm, making an **exact copy** of the DNA on the other strand. This happens with the help of enzymes. The result is **two molecules** of DNA **identical** to the **original molecule** of DNA:

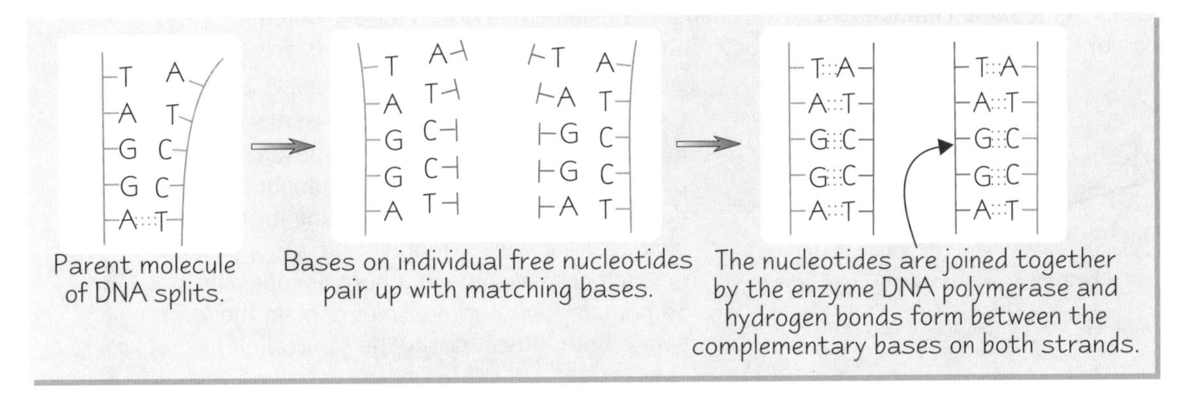

2) This type of copying is called **semi-conservative replication** — because **half** of the new strands of DNA are from the **original** piece of DNA.

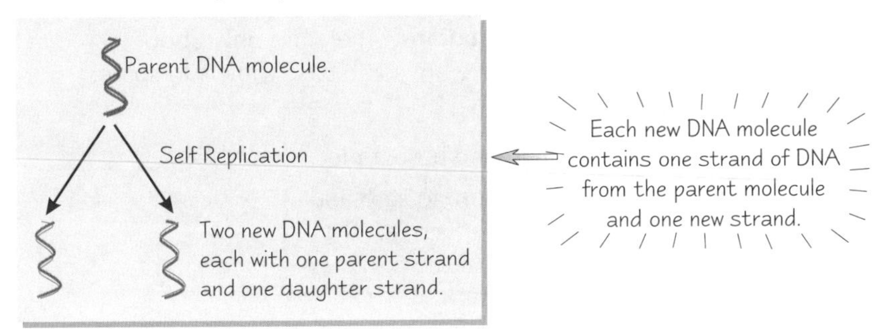

Practice Questions

Q1 What three things are nucleotides made from?

Q2 Which base pairs join together in a DNA molecule?

Q3 What type of bonds join the bases together?

Q4 What is the name used to describe the type of replication in DNA?

Exam Questions

Q1 Explain how the structure of DNA is related to its function. [2 marks]

Q2 Describe, using diagrams where appropriate, how nucleotides join together and how two single strands of DNA become joined. [5 marks]

Q3 Describe and explain the semi-conservative method of DNA replication. [6 marks]

Give me a D, give me an N, give me an A! What do you get? — very confused...

You need to know the basic structure of DNA and RNA and also how DNA's structure makes it good at its job. Then there's self replication to get to grips with — hmmm, rather you than me. I'm afraid there's nowt else you can do except buckle down, pull your socks up and get all them facts learnt.

The Genetic Code

Here comes some truly essential stuff — the genetic code is the real nitty-gritty of biology. So eyes down for some serious fact-learning. I'm afraid it's all horribly complicated — all I can do is keep apologising. Sorry.

DNA Contains the **Basis** of the **Genetic Code**

Genes contain the genetic information that determines the development of all organisms. **Genes** are sections of DNA that code for a specific **sequence of amino acids** that forms a particular **protein**. The way that DNA codes for proteins is called the **genetic code**.

A gene can exist in more than one form. These forms are called **alleles** — they code for **different types** of the **same characteristic**. For example, the gene that codes for **eye colour** exists as one of two alleles — one codes for the colour **blue** and the other codes for **brown.**

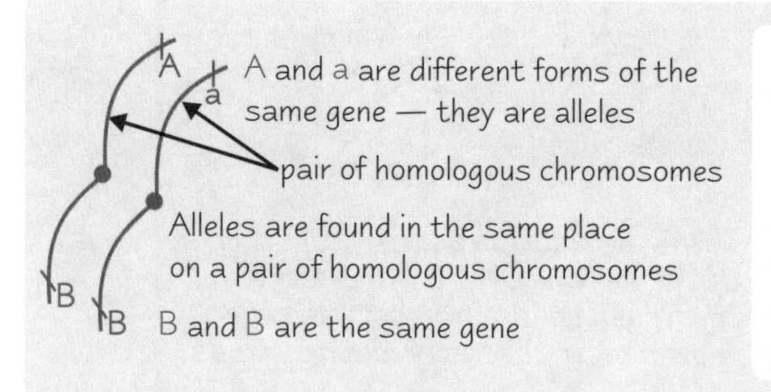

Genes are located along **chromosomes** found in the nucleus of cells. These chromosomes are paired into matching pairs (**homologous pairs**) during cell division. Alleles coding for the same characteristic will be found at the **same position** on each chromosome in a homologous pair. So homologous chromosomes contain the same genes, but not necessarily the same alleles.

1) Genes code for specific amino acids with sequences of three bases, called **base triplets**. Different sequences of bases code for different amino acids. For example AGA codes for serine and CAG codes for valine.
2) There are **64** possible **base triplet combinations**. There are only about **20** amino acids in human proteins so there are some base triplets to spare. These aren't wasted though:
   - some amino acids use more than one base triplet.
   - some base triplets act as 'punctuation' to stop and start production of an amino acid sequence. These create **stop codons** and **start codons** (see p.42).

The Genetic Code is **Non-Overlapping** and **Degenerate**

1) In the genetic code, each base triplet is read in sequence, separate from the triplet before it and after it. Base triplets **don't share** their **bases** — so the code is described as **non-overlapping**.
2) The genetic code in DNA is also described as **degenerate**. This is because there are **more triplet codes** than there are amino acids. Some **amino acids** are coded for by **more than one base triplet**, e.g. tyrosine can be coded for by TAT or TAC.
3) There are sections of DNA that **don't code** for amino acids — these lengths of DNA are called **introns** (all the bits that do form part of the genetic code are called **exons**). Introns are removed from DNA during protein synthesis. Their purpose isn't known for sure.

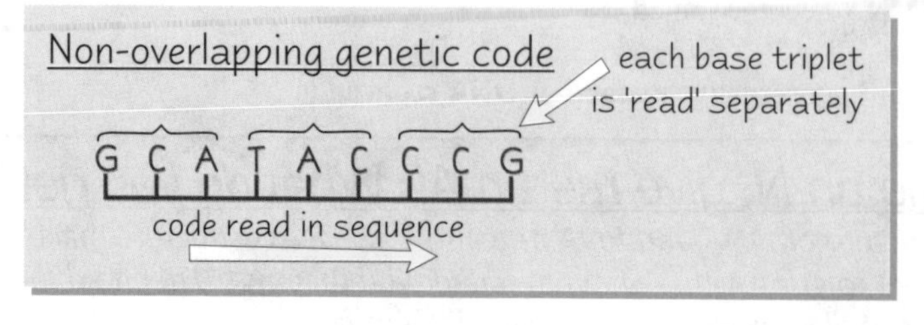

Types of RNA

There are Three Types of RNA

There are **three types** of RNA — **ribosomal RNA** (rRNA), which forms part of **ribosomes**, and **messenger RNA** (mRNA) and **transfer RNA** (tRNA), both which are involved in **making proteins** (see over the page for more on this — bet you can't wait). You need to know details of the **structures** of **mRNA** and **tRNA**.

Messenger RNA (mRNA)

1) **mRNA** is a **single polynucleotide strand** that's formed in the **nucleus**.
2) The important thing to know about it is that it's formed by using a section of a **single strand of DNA** as a **template**. **Specific base pairing** means that mRNA ends up being an exact **reverse copy** of the DNA template section (see the piccy on the right to make sense of this).
3) You also need to know that the **3 bases in mRNA** that pair up with a base triplet on the DNA strand are called a **codon**. Codons are dead important for making proteins (see p.42), so **remember this word**. Make sure you realise that a codon has the **opposite bases** to a base triplet (except the base **T** is replaced by **U** in **RNA**).

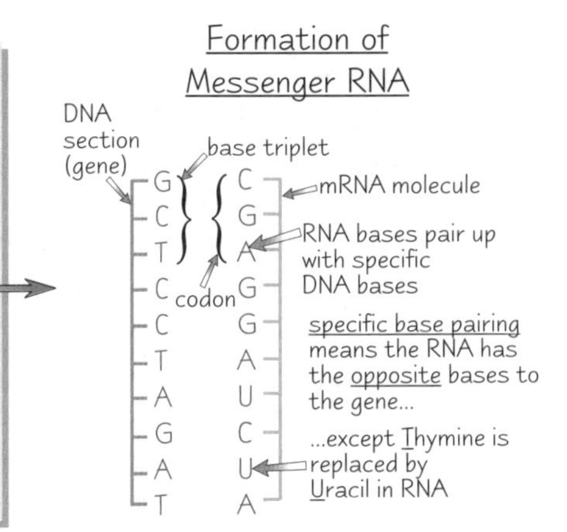

Transfer RNA (tRNA)

1) **tRNA** is a **single polynucleotide strand** that's folded into a **clover shaped molecule**.
2) Each tRNA molecule has a **binding site** at one end, where a specific **amino acid** attaches itself to the bases there.
3) Each tRNA molecule also has a specific sequence of **three bases** at one end of it, called an **anticodon**.
4) The significance of binding sites and anticodons are all revealed over the page. But you need to know **where they are found** on a tRNA molecule, so learn the diagram on the left off by heart.

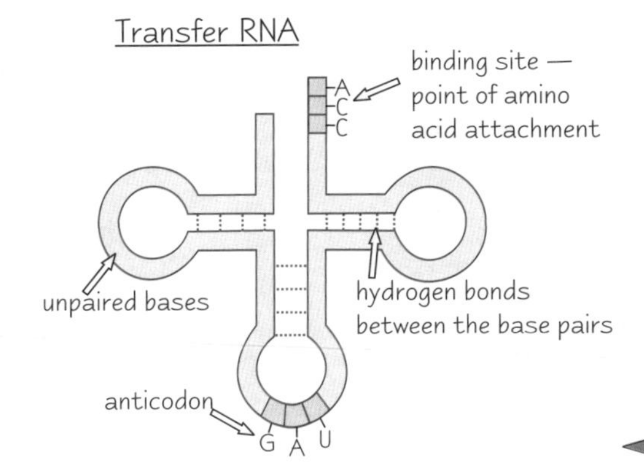

Practice Questions

Q1 What is an allele?

Q2 Explain why the genetic code is described as being "degenerate".

Q3 What three types of RNA are there?

Q4 What is the name of the group of three bases on mRNA that correspond to a base triplet on DNA?

Q5 What shape does a chain of tRNA fold itself into?

Exam Questions

Q1 Write a definition of a gene. [2 marks]

Q2 A gene has the following base sequence: AATGCAGGCTCT.
Write the base sequence of the mRNA molecule that's formed along this gene in the nucleus. [2 marks]

My genes are degenerate — there's a hole in the back pocket... *(I'll get my code)*

Quite a few terms to learn here — you're on the inescapable road to science geekville I'm afraid, and it's a road lined with crazy diagrams and strange words. The genetic code is sooo important — so make sure you understand what's going on. You need to learn the structure of mRNA and tRNA — it'll help you understand protein synthesis, on the next page.

Protein Synthesis

You've learnt about the genetic code and types of RNA — now you can learn all about their role in making proteins. This stuff is biology at its most clever, and it's probably going on inside you right now. Weird.

Protein synthesis (making proteins) happens in **two stages** — **transcription** and **translation**. It involves both DNA and RNA.

*First Stage — **Transcription** Occurs in the **Nucleus***

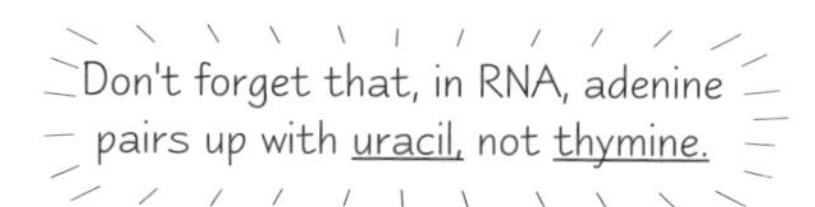

In **transcription** a '**negative copy**' of a **gene** is made. This copy is called **mRNA**.

1) A gene (a section of DNA) in the DNA molecule **uncoils** and the hydrogen bonds between the two strands in that section break, separating the strands.
2) One of the strands is then used as the **template** for transcription — it's called the '**sense strand**'.
3) Free **RNA nucleotides** in the nucleus line up alongside the template strand. Once the RNA nucleotides have **paired up** with their **complementary bases** on the DNA strand they're joined together by the enzyme **RNA polymerase**.
4) The strand that's formed is **mRNA**.
5) It then moves out of the nucleus through a nuclear pore, and attaches to a **ribosome** in the cytoplasm, where the next stage of protein synthesis takes place.
6) When enough mRNA has been produced, the uncoiled strands of DNA re-form the hydrogen bonds and **coil back into a double helix**, unaltered.

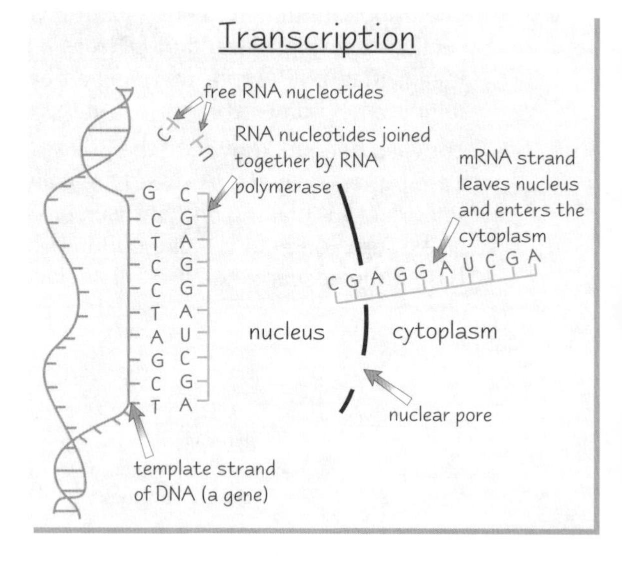

*Second Stage — **Translation** Occurs in a **Ribosome***

In **translation**, **amino acids** are stuck together to make a **protein**, following the order of amino acids coded for on the mRNA strand.

1) The **mRNA strand** has travelled to a ribosome in the cytoplasm, and attached itself.
2) All 20 **amino acids** needed to make human proteins are in the cytoplasm. tRNA molecules attach to the amino acids and transport them to the ribosome.
3) In the ribosome, a tRNA molecule binds to the start of the mRNA strand. This tRNA molecule has the **complementary anticodon** to the **first codon** on the mRNA strand, and attaches by **base pairing**. Then a second tRNA molecule attaches itself to the **next codon** on the mRNA strand in the **same way**.
4) The two amino acids attached to the tRNA molecules are joined together with a **peptide bond** (using ATP and an enzyme).
5) The first tRNA molecule then **moves away** from the ribosome, leaving its amino acid behind. The mRNA then **moves across** the ribosome by one codon and a third tRNA molecule binds to the **next codon** that enters the ribosome.
6) This process continues until there's a **stop codon** on the mRNA strand that doesn't code for any amino acid. You're left with a line of amino acids joined by peptide bonds. This is a **polypeptide chain** — the **primary structure** of a protein. The polypeptide chain moves away from the ribosome and translation is complete.

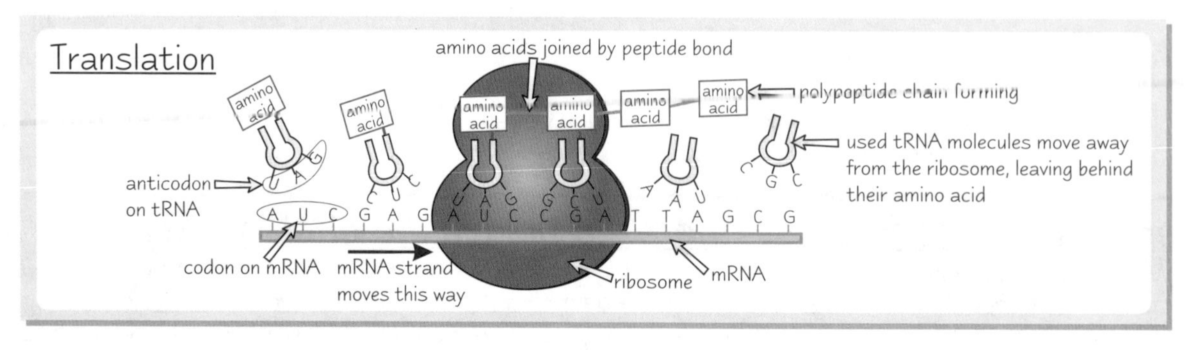

Protein Synthesis

At the End of Protein Synthesis, Proteins are Formed

So, in protein synthesis, the **sequence of codons** on the mRNA strand determines the **sequence of amino acids** that makes up the primary structure of the protein.
When **translation** is complete, the polypeptide chain that it creates folds itself into its **secondary** and **tertiary** structure, and a **protein** is formed (see p.12). These proteins are then used for all kinds of vital growth and development functions in the body.

The Structure of Enzymes is Determined by Protein Synthesis

1) All **enzymes** are **proteins**, which are sequences of amino acids.
2) The amino acid sequences are determined by the base sequence in DNA, so **DNA** determines the **structure of enzymes**.
3) Enzymes speed up all our **metabolic pathways** (see p.18).
They have a big influence over how our **genes** are **expressed physically**, by controlling chemical reactions required for growth and development. This physical expression of the gene contributes to the organism's **phenotype** (what the organism looks like).

This flowchart shows how the **genetic code** and **protein synthesis** determine our **metabolic processes** and therefore our **phenotype**:

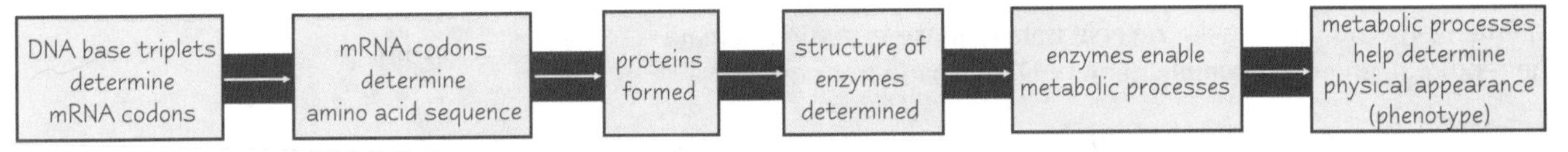

Practice Questions

Q1 What are the two main stages in protein synthesis?

Q2 Which base is not present in RNA?

Q3 Write the RNA sequence which would be complementary to the following DNA sequence.
AATTGCGCCCG

Q4 Where does transcription take place?

Q5 Where does translation take place?

Exam Questions

Q1 Explain the terms codon and anticodon. [2 marks]

Q2 Describe the process of protein synthesis. [10 marks]

mRNA codons join to tRNA anticodons?! — I need a translation please...

When you first go through protein synthesis it might make approximately no sense, but I promise its bark is worse than its bite. All those strange words disguise what is really quite a straightforward process — and the diagrams are dead handy for getting to grips with it. Keep drawing them yourself, 'til you can reproduce them perfectly.

Recombinant DNA

Genetic engineering is a dead popular exam topic because it shows how biology relates to real life — and examiners love all that stuff. These pages explain the process for manufacturing genes and, you've got to admit, it's kinda cool.

Recombinant DNA is like 'Home-Made' DNA

DNA has the **same** structure of **nucleotides** in **all organisms**. This means you can **join together** a piece of DNA from one organism and a piece of DNA from another organism.

DNA that has been **genetically engineered** to contain DNA from another organism is called **recombinant DNA**. It has **useful applications**. In the example below, a **human gene** coding for a **useful protein** is inserted into a bacterium's DNA. When the **bacterium reproduces**, the **gene** is **reproduced** too. The gene is **expressed** in the bacterium, and so the bacterium **produces** the **protein** which is coded for by the gene.

First You Need to Isolate the Useful Gene

1) First you **find** the gene you want in the donor cell. Then the **DNA** containing the gene is **removed** from the cell.

2) Next you **cut out** the **useful gene** from the DNA using **restriction endonuclease** enzymes. These leave a **sticky end** (tail of unpaired bases) at each end of the useful gene. This stage is called **restriction**.

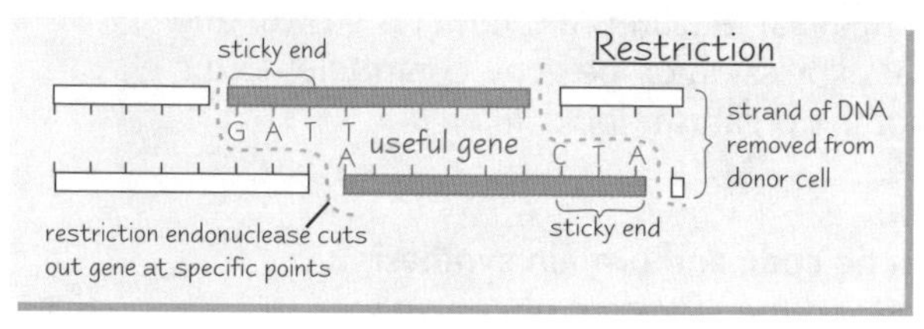

Alternatively, you can use the **reverse transcriptase** enzyme to create the required gene from **complementary DNA**, made from mRNA:

Transcription is when mRNA is made from DNA (see p.42). **Reverse transcriptase** is a nifty enzyme that runs this process backwards and makes DNA from mRNA. It's useful in genetic engineering because **cells** that make **specific proteins** usually contain **more mRNA molecules than genes**. For example, **pancreatic cells** produce the protein **insulin**. They have loads of mRNA molecules that make insulin — but only two copies of the DNA gene for insulin.

1) **mRNA** is **extracted** from donor cells.
2) The mRNA is mixed with **free DNA nucleotides** and **reverse transcriptase**. The reverse transcriptase uses the mRNA as a **template** to synthesise a new strand of **complementary DNA**.
3) The complementary DNA can be made **double-stranded** by mixing it with DNA nucleotides and **polymerase enzymes**. Then the useful gene from the double-stranded DNA is inserted into a plasmid (see next page) so the bacteria can make lots of the product of the gene.
4) The diagram shows a monkey swinging from a strand of DNA. Try to ignore him — he's attention-seeking.

Then You Need to Prepare the Vector

Then you need to **prepare** the other bit of DNA that you're **joining** the useful gene to (called a **vector**). The main vectors used are **plasmids** — these are small, **circular molecules** of DNA found in **bacteria**. ⟹ They're useful as vectors because they can **replicate** without interfering with the bacterium's own DNA (see next page for more on vectors). To prepare it, the same **restriction enzyme** is used to **cut out** a section of the plasmid. The **sticky ends** that are left have bases that are **complementary** to the bases on the sticky ends of the **useful gene**.

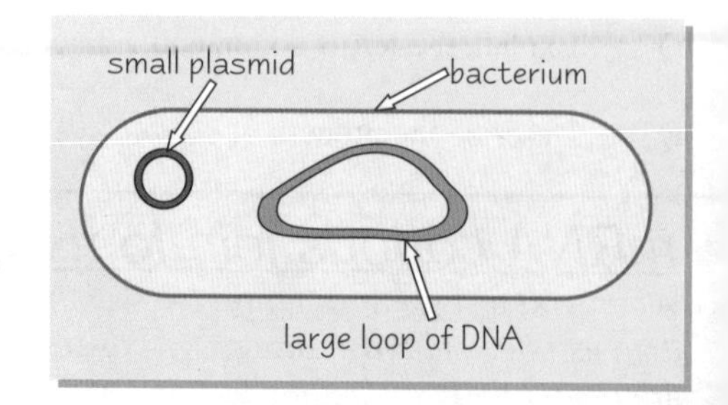

Recombinant DNA

Then You Need to Transfer the Useful Gene to the Vector

Next, you need to **join** the **isolated useful gene** to the **plasmid vector DNA** — this process is called **ligation**. This is where the **sticky ends** come in — **hydrogen bonds** form between the **complementary bases** of the sticky ends. The DNA molecules are then 'tied' together with the enzyme, **ligase**. The new DNA is called **recombinant DNA**.

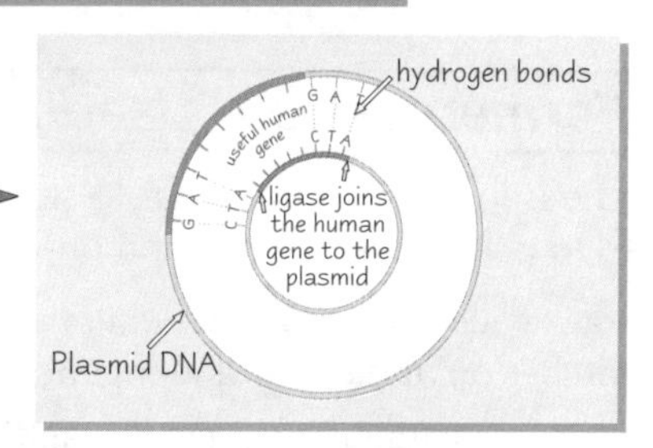

Finally you Need to Move the Vector into the Micro-organism

The **vector** carries recombinant DNA containing the useful gene into a **microbe** (usually a bacterium). Vectors are usually **plasmids** (see the above example) or **viruses**, like the **λ** (**Lambda**) **phage virus**.

1) **Host bacteria** have to be **persuaded** to take up **plasmid** vectors. They're placed into cold **calcium chloride** solution to make their cell membranes more **permeable**. Then the plasmids are added and the mixture warmed up. The **sudden cold** causes some of the bacteria to take up the plasmids.
2) When a **phage** is used as a vector, the phage DNA is combined with the useful gene to make recombinant DNA. The phage then **infects** a bacterium by injecting its DNA strand into it. The phage DNA is then **integrated** into the bacterium's DNA.
3) Bacteria that have taken up vectors are said to be **transformed**.

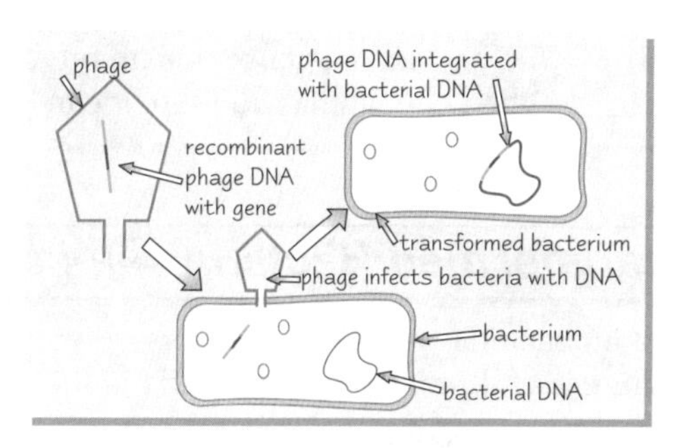

The useful gene is now in the **recombinant DNA** of a **plasmid vector** in a **bacterium**. So now the bacterium is made to **reproduce loads**, so you get loads of **copies** of the gene (see next page for how this happens). Clever, huh.

Practice Questions

Q1 What is recombinant DNA?

Q2 What do restriction endonuclease enzymes do?

Q3 How does reverse transcriptase work?

Q4 How are useful genes joined to plasmids?

Q5 What is a vector? Give two examples.

Exam Questions

Q1 Plant breeders have found a variety of cabbage that is resistant to a pest called root fly. This is because the cabbage produces a protein that inhibits one of the fly's digestive enzymes. Plant breeders want to manufacture this protein so they can insert it into carrots too. Describe how scientists could:

a) remove the gene that produces the inhibitor from the cabbage. [2 marks]

b) insert this gene into the DNA of a bacterium. [2 marks]

Q2 Explain how a virus could be used as a vector. [3 marks]

Monkey vectors — Plasmid of the Apes...

You see, biology isn't just an evil conspiracy to keep students busy — it has loads of really important uses in real life. For example, sufferers of genetic diseases now have a far greater chance of having successful treatment, because you can make many copies of the healthy gene that they lack. Round of applause for biology, that's what I say.

Recombinant DNA

Having a single copy of a useful gene is all very well, but thousands of copies are even better. Fortunately, we can use microbes as micro-factories. They can copy the gene and make the gene product, but only if you treat 'em right.

Industrial Fermenters are Used for Large Scale *Protein Production*

Once a useful gene is in a bacterium, it has to start working to produce the **protein**. An **industrial fermenter** is used to culture the bacteria and produce a **large amount** of the gene product (the protein).

1) A **promoter** gene is often included along with the **donor** gene when the **recombinant DNA** is made. This '**switches on**' the useful donor gene so it starts making the **protein**.
2) The **bacteria** containing the recombinant DNA are grown or cultured in a **fermenter**. Inside the fermenter, they're given the **ideal conditions** needed for rapid growth (see p.32 for more on industrial fermenters). They **reproduce quickly**, until there are millions of bacteria inside the fermenter.
3) The **plasmids**, including plasmids made of recombinant DNA, **replicate** at each cell division — so each new bacterium contains the useful gene.
4) As the bacteria grow, they start producing the **human protein** that the donor gene in the plasmid codes for, e.g. **human factor VIII** or **human insulin**.
5) The bacteria can't use human protein, so it **builds up** in the medium inside the fermenter. When enough has built up, it can be **extracted** and processed for use.

Transformed Bacteria can be Identified Using *Marker Genes*

Not all bacteria will take up vectors, so you need a way to **pick out** the **transformed** bacteria, so that you only culture **useful**, **transformed bacteria** in the fermenter. The easiest way is to use **antibiotic marker genes**.

1) When recombinant DNA is produced in plasmids, a marker gene for **antibiotic resistance** is also inserted into the plasmid, along with the donor gene. This means that transformed bacteria contain both the donor gene **and** the gene for antibiotic resistance.
2) After being mixed with plasmids, the bacteria are cultured on an agar plate called the **master plate**.
3) Once bacteria have grown on the master plate, **replica plating** is used to isolate **transformed bacteria**:

- A **sterile velvet pad** is pressed onto the master plate. This picks up some bacteria from each colony.
- The pad is pressed onto a fresh agar plate, **containing an antibiotic**. Some of the bacteria from each colony are transferred onto the agar surface.
- Only transformed bacteria can grow and reproduce on the replica plate — the others don't contain the antibiotic-resistant gene, so they stop growing.

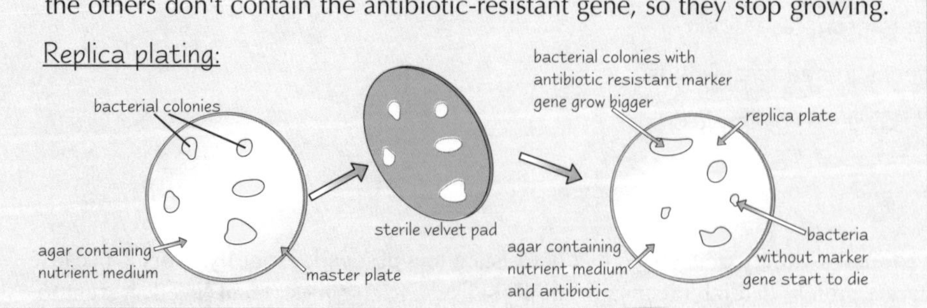

Genetic Engineering has *Important Medical Uses*

Sometimes humans can't produce a certain **protein** because the **gene** that codes for it is **faulty**. By creating recombinant DNA, genes can be artificially **manufactured**, to replace the faulty genes. This has important **medical benefits**. For example:

1) **Diabetics** don't have the healthy gene needed to make the protein **insulin**, which controls blood sugar levels.
2) **Haemophiliacs** lack the healthy gene that codes for the protein, **factor VIII**, which allows blood to clot.
3) Genetic engineering allows these genes to be manufactured in bacterial cells and used to help sufferers.

The Ethics of Genetic Engineering

Genetic Engineering Raises *Ethical* and *Moral* Issues

Genetic engineering means that we now have the technology to manipulate genes.
But there are **ethical** and **moral** issues to consider.

Potential Benefits of Genetic Engineering	Concerns over Genetic Engineering
Specific medicines could be developed to treat diseases. A current example of this is the production of genetically engineered human insulin to control diabetes.	Risk of foreign genes entering **non-target** organisms and disrupting functions.
Faulty genes could be identified and replaced, preventing genetically inherited diseases.	Risk of **accidental transfer** of unwanted genes, which could damage the recipient.
Parents could make sure their babies didn't have faulty genes before they were born.	Doctors might be under pressure to implant embryos that could provide **transplant material for siblings**. Those who can afford it might decide which characteristics they wish their children to have (**designer babies**), creating a 'genetic underclass'. The **evolutionary consequences** of genetic engineering are unknown. **Religious concerns** about 'playing God'.
Crops could be engineered to give **increased yields** or exploit new habitats	**Health concerns** over human consumption of genetically modified foods.

Practice Questions

Q1 Explain how a useful protein is manufactured on a large scale.

Q2 How can you use replica plating to isolate transformed bacteria?

Q3 What proteins do diabetics and haemophiliacs lack?

Q4 Give three ethical concerns about genetic engineering.

Q5 Give three potential benefits of genetic engineering.

Exam Questions

Q1 A human gene was combined with a plasmid which also contained a gene coding for resistance to ampicillin (an antibiotic).

The plasmid was added to a bacterial culture.

a) Which technique could be used to find out whether the bacteria had taken up the plasmid containing the gene? [1 mark]

b) Explain how this technique works. [5 marks]

Q2 Evaluate the advantages and disadvantages of genetic engineering. [6 marks]

Designer babies — mine's Gucci, daahling...

The antibiotic resistance gene is the genetic marker that you need to know about — learn how it's used to identify transformed bacteria, including how replica plating isolates the transformed bacteria. Then ponder for a while, with an earnest, serious look, the ethical issues surrounding genetic engineering. Hmmmmmmmm.

Immunology

The immune system protects the body from pathogens (organisms that cause disease). It helps the body recognise them as foreign, and destroys them with a mixture of funny-sounding cells, like phagocytes and lymphocytes. Hurrah.

Antigens and Antibodies know What Should and Shouldn't be in our Body

Antigens and **antibodies** work together to let our cells distinguish 'self' from 'non-self' — they know when something is in our body that shouldn't be.

1) **Antigens** are large, organic molecules found on the surface of **cell membranes**. Every individual has their own **unique** cell surface molecules, which the immune system recognises. So when a **pathogen** like a bacterium invades the body, the antigens on its cell surface are identified as **foreign** by the immune system.
2) When the body detects foreign antigens, it makes **antibodies** — **protein molecules** that bind to specific antigens, giving an **antigen-antibody complex**.
3) Antibodies have **two binding sites**, so they can attach to two antigen molecules. They deal with pathogens by **clumping** or **linking** them together to make it easier for them to be engulfed by **phagocytes**. They can also **rupture** foreign cells, (which kills them) and inactivate any **toxins** they produce.

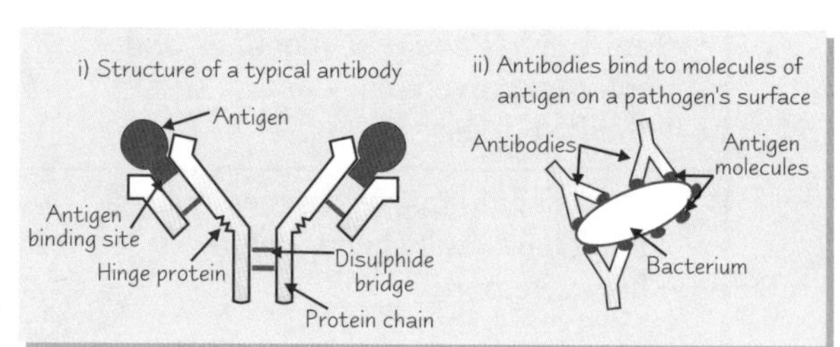

B-lymphocytes are Cells that Produce Antibodies to Fight Disease

Lymphocytes are white blood cells with **glycoproteins** on their surface to **recognise foreign antigens**. The two main types are **B-lymphocytes** that mature in the bone marrow, and **T-lymphocytes** that mature in the thymus gland. You only need to know about the role that **B-lymphocytes** (**B-cells**) play in our immune response.

1) **B-cells** produce and release antibodies. Each **B-cell** has a **specific antibody** on its membrane.
2) When a B-cell meets a complementary foreign **antigen**, the antigen binds to the antibody's **receptor site**. This stimulates the B-cell to release the antibody into the blood.
3) The B-cell then divides by **mitosis**, producing many **clones**, called **plasma cells**. These cells release large quantities of the same antibody, hopefully killing the pathogen.
4) Most of the plasma cells then die, but some become **memory B-cells**, which provide us with immunity against a pathogen if it enters the body again. There's more on memory cells below.

NB — We produce an almost infinite number of B-lymphocytes, each with a different antibody on its membrane. Only the ones that bind to their matching antigen go on to reproduce — the rest just die if they're not used.

The Immune Response for Antigens can be Memorised

The **first time** a particular **pathogen enters** the body the **immune response is slow**, because there aren't many B-cells that can make the antibody needed to bind to it. This is the **primary response**. Eventually the body will probably produce enough of the antibody to overcome the infection, but meanwhile the infected person will show **symptoms** of the disease. BUT, after being exposed to an antigen, B-cells produce **memory cells** that record the '**recipe**' **for antibodies** to fight them.

So if the **same pathogen** enters the body again, these **memory cells** can produce the right cells and antibodies for fighting the pathogen **very quickly**. This is known as the **secondary response**. It often gets rid of the pathogen before you begin to show any symptoms.

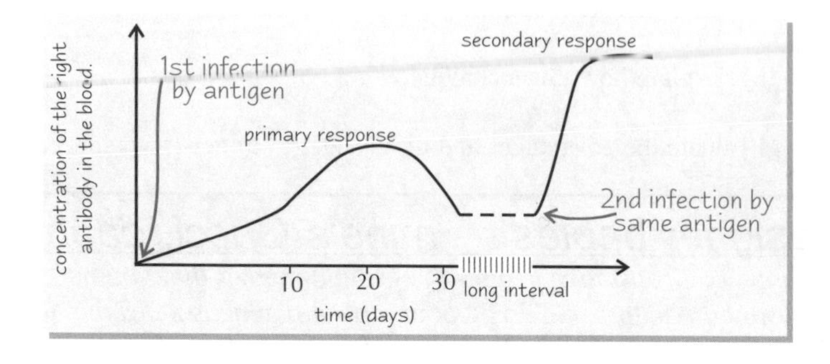

Blood Groups

Blood, blood, lovely blood (apologies to blood-phobics for the unnecessary repetition). We all belong to a blood group, and which one it is depends on the antigens and antibodies found in our blood. I'm AB personally. Bzzzzzz.

There are Four Blood Groups — A, B, AB and O

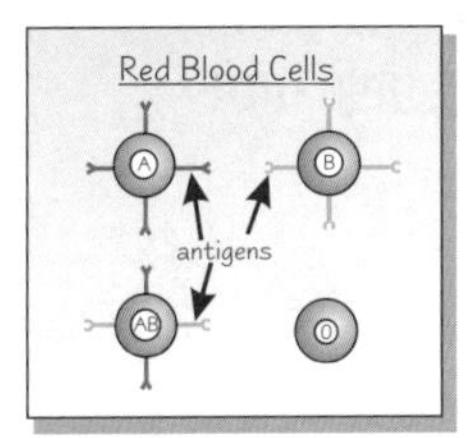

1) **Red blood cells** carry special **antigens** (called **agglutinogens**) on their surface. There are two types of red blood cell antigen — **A** and **B**.
2) These determine **blood groups** — a person with type A antigens will be blood group A, a person with type A **and** B antigens = blood group AB and so on... If you don't have any antigens on your red blood cells you're blood group O.

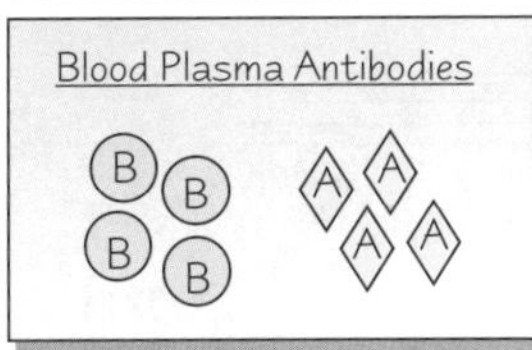

1) **Blood plasma** carries special **antibodies** (called **agglutinins**). Again there are two types — A and B.
2) If blood antigens and antibodies of the **same type** mix, **agglutination** occurs — the agglutinins bind to the agglutinogens on the blood cells, forming a **clump**. This can be **fatal** because the clumps can **block arteries**.
3) So we **don't** have agglutinogens and agglutinins of the **same type** in our bodies. If we have type A agglutinogens, we'll have type B agglutinins and so on...

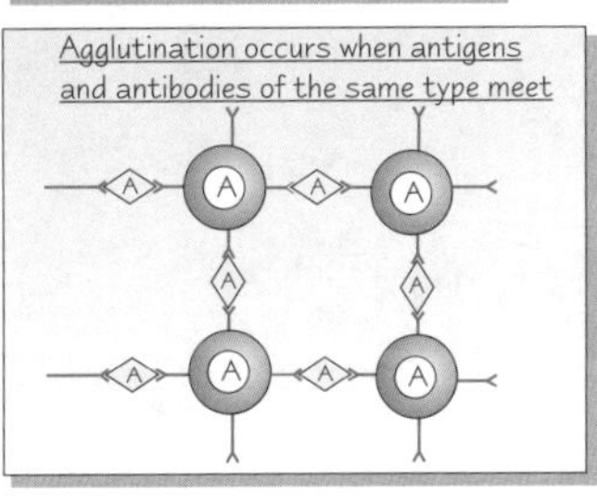

Blood groups	Red Blood Cell Antigens	Blood Plasma Antibodies
A	A	B
B	B	A
AB	A+B	none
0	none	A+B

Most **blood transfusions** have to be from someone of the same blood group to prevent **agglutination** happening in the transfusion patient's blood. But there are two exceptions:

People with **blood group O** (with **no agglutinogens** on their red blood cells) can **donate** blood to anyone. Also, people with **blood group AB** (with **no agglutinins** in their blood plasma) can **receive** blood from anyone. The reason for this is that it doesn't usually matter if donated blood contains the wrong **agglutinins**, because there won't be a high enough concentration of them to cause serious agglutination. BUT, donated blood can't contain the wrong **agglutinogens**, because the high concentration of **agglutinins** in the patient's blood will attack them, causing fatal agglutination.

Practice Questions

Q1 What are antigens?

Q2 What is an antibody?

Q3 Where do B-lymphocytes come from, and what is their function?

Q4 What is the difference between an agglutinogen and an agglutinin?

Q5 Describe what "agglutination" is.

Exam Questions

Q1 How do B-lymphocyte cells protect the body from disease? [6 marks]

Q2 Explain what would happen if a person of Blood Group A received a transfusion of Blood Group B. Why could this be life-threatening? [7 marks]

My Uncle's worried — my Anti-Gen is obsessed with bodies...

It has to be said, the names of some of the cells are verging on the ridiculous. I'm sure we're all very grateful to these phagocytes and B-lymphocytes for doing their bit to defend us from the attacking hordes. But why couldn't they have names like Bill, Jim and Dave? And as for agglutinins and agglutinogens — don't even get me started...

Genetic Fingerprinting

The tiniest bit of DNA is enough to identify you. You're shedding hairs and flakes of skin-scum all the time (nice) — it's like leaving a personal trail of, well, ***you*** *wherever you go. Scary (and quite grim) thought.*

You can **Identify People** from their **DNA** by **Cutting** it into **Fragments**

The DNA from cells in a sample of blood, sweat, semen, skin or hair can be used to **identify a person**, if the sample is big enough. This is done by using **enzymes** to cut the DNA up into **fragments**, then looking at the **pattern** of fragments, which is **different** for everyone (except identical twins). This is called a person's **genetic fingerprint**.

1) To **cut up** the DNA into DNA fragments you add specific **restriction endonuclease** enzymes to the DNA sample — each one **cuts** the DNA every time a **specific base sequence** occurs. **Where** these base sequences occur on the DNA **varies** between everyone (except identical twins), so the number and length of DNA fragments will be different for everyone.
2) Next you use the process of **electrophoresis** to separate out the DNA fragments by size:

How Electrophoresis Works:

1) The DNA fragments are put into **wells** in a slab of **gel**. The gel is covered in a **buffer solution** that **conducts electricity**.
2) An **electrical current** is passed through the gel. DNA fragments are **negatively charged**, so they move towards the positive electrode. **Small** fragments move **faster** than large ones, so they **travel furthest** through the gel.
3) By the time the current is switched off, all the fragments of DNA are **well separated**.

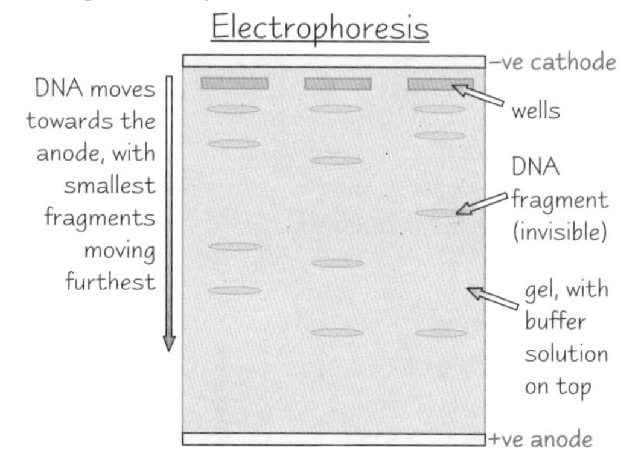

In electrophoresis the DNA fragments **aren't visible** to the eye — you have to do something else to them before you can **see their pattern**. Coincidentally that's what the next section's about...

Gene Probes Make the **Invisible** 'Genetic Fingerprint' **Visible**

DNA fragments separated by electrophoresis are invisible.
A radioactive DNA probe (also called **gene probe**) is used to show them up:

1) A **nylon membrane** is placed over the electrophoresis gel, and the DNA fragments **bind** to it.
2) The DNA fragments on the nylon membrane are **heated** to separate them into **single strands**.
3) **Radioactive gene probes** are then put onto the nylon membrane. (It's the **phosphorus** in the gene probes' sugar-phosphate backbones that's radioactive.) The probes are warmed and **incubated** for a while so that they'll attach to any bits of complementary DNA in the DNA fragments.

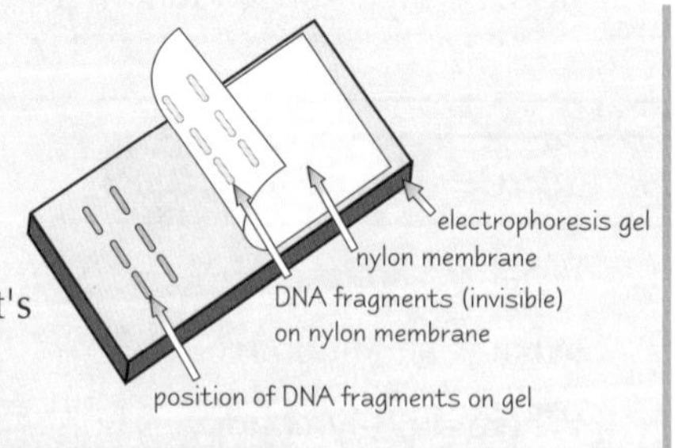

4) The nylon membrane is then put on top of unexposed **photographic film**. The film goes **dark** where the radioactive gene probes are **present**, which **reveals the position** of the **DNA fragments**. The pattern is different for every human — it's **unique** like a fingerprint.

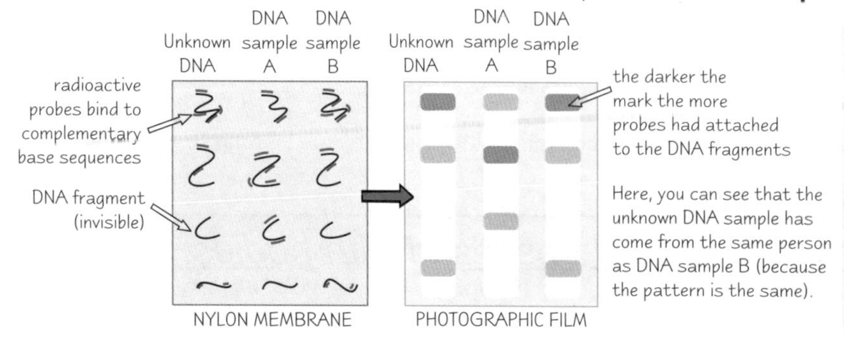

Genetic fingerprinting is incredibly useful. **Forensic investigations** use it to confirm the identity of suspects from blood, hair, skin, sweat or semen samples left at a crime scene, or to establish the identity of victims. **Medical investigations** use the same technique for **tissue typing, paternity tests** and **infection diagnosis**.

Polymerase Chain Reaction

The **Polymerase Chain Reaction** (PCR) Creates Millions of **Copies** of DNA

Some samples of DNA are too small to analyse. The **Polymerase Chain Reaction** makes millions of copies of the smallest sample of DNA in a few hours. This **amplifies** DNA, so analysis can be done on it. PCR has **several stages**:

1) The DNA sample is **heated** at **95°C**. This breaks the hydrogen bonds between the bases on each strand. But it **doesn't** break the bonds between the ribose of one nucleotide and the phosphate on the next — so the DNA molecule is broken into **separate strands** but doesn't completely fall apart.
2) **Primers** (short pieces of DNA) are attached to both strands of the DNA — these will tell the **enzyme** where to **start copying** later in the process. They also stop the two DNA strands from joining together again.

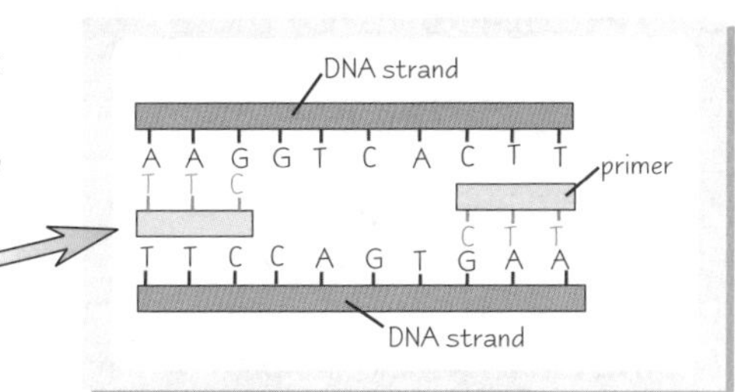

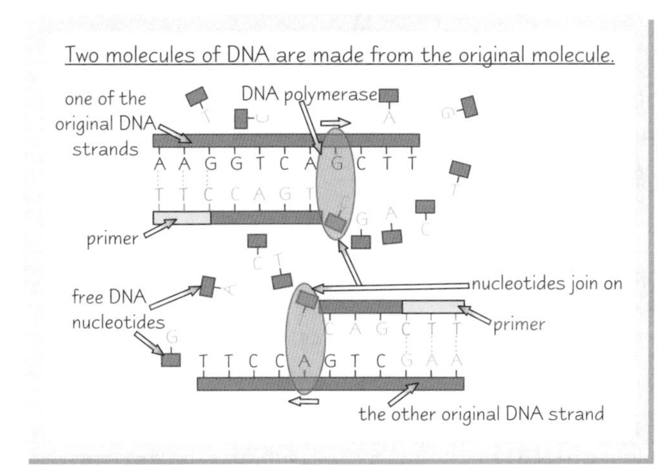

3) The DNA and primer mixture are **cooled** to **40°C** so that the primers can **fully bind on** to the DNA.
4) Free **DNA nucleotides** and the enzyme **DNA polymerase** are added to the reaction mixture. The mixture is heated to **70°C**. Each of the original DNA strands is used as a **template**. Free DNA nucleotides pair with their complementary bases on the template strands. The DNA polymerase attaches the new nucleotides together into a strand, starting at the primers.
5) The cycle starts again, using **both** molecules of DNA. Each cycle **doubles** the amount of DNA.

Practice Questions

Q1 Give the full name of the technique used to increase the amount of DNA from a very small sample.

Q2 Name the enzyme added to DNA before gel electrophoresis.

Q3 Which part of a gene probe is radioactive?

Q4 State three uses of genetic fingerprinting in medical investigations.

Q5 Name the enzyme used in PCR.

Exam Question

Q1 Police have found incriminating DNA samples at the scene of a murder.
They have a suspect in mind, and want to prove that the suspect is guilty.

a) Name the technique that the police could use to confirm the guilt of the suspect. [1 mark]

b) Explain how this technique would be carried out. [5 marks]

You see that sweat stain there — that's you, that is...

Genetic fingerprinting has revolutionised medicine and forensic science. Remember that the PCR amplifies small DNA samples so genetic analysis can be done. You need to learn the PCR process, plus the process of electrophoresis and how it's involved in genetic fingerprinting. Also, learn what genetic fingerprinting is used for in the real world.

Adaptations of Cereals

Cereal plants are really important in our diet. They provide ***carbohydrate*** *for* ***energy****.*
So grab yourself a quick bowl of cornflakes, and get on with learning all these lovely facts.

Cereals have different **structural** and **physiological adaptations** so they can grow in **different environments**.

Rice is Adapted to Grow in Swampy Conditions

Rice grows in the **swampy paddy fields** of Asia so it needs to be able to survive in **water**.

1) There are **no air spaces** in the waterlogged soil, so oxygen can't get to the roots through the soil. Most plants would **die** in these conditions because they need oxygen for **aerobic respiration**.
2) **Structurally**, rice is adapted by having a special tissue called **aerenchyma** in its leaves, stems and roots, which has large air spaces in it. The air spaces help the plant 'float' in the waterlogged soil and allow oxygen to diffuse from the air to the roots.
3) **Physiologically**, rice is adapted by being able to use **anaerobic respiration** to provide energy to the roots. It can **tolerate large quantities of ethanol**, a product of anaerobic respiration that's poisonous to most plants.

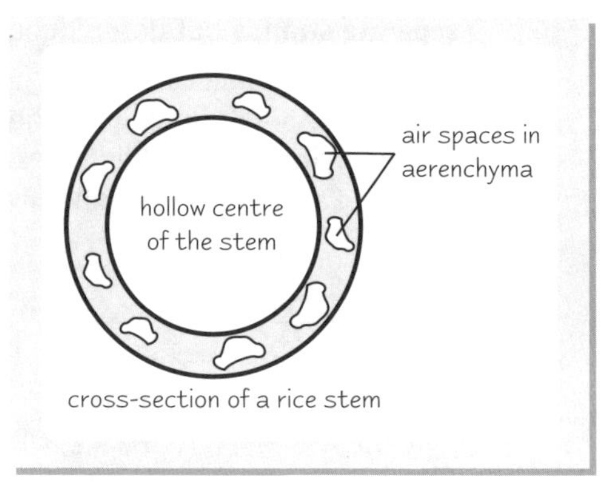

Maize Uses a Special Method of Photosynthesis

Maize is a tropical plant that grows in **hot places** with **high light intensity**. These conditions affect **photosynthesis**:

1) **High temperatures** increase **water loss** by transpiration, which closes the stomata.
2) **Closed stomata** mean carbon dioxide can't enter the plant, leading to **low carbon dioxide levels** in leaves.
3) The **closed stomata** also mean **oxygen builds up** inside leaves. The built-up oxygen **inhibits the enzyme** that's used to '**fix**' carbon dioxide into the form that's used in **photosynthesis**.

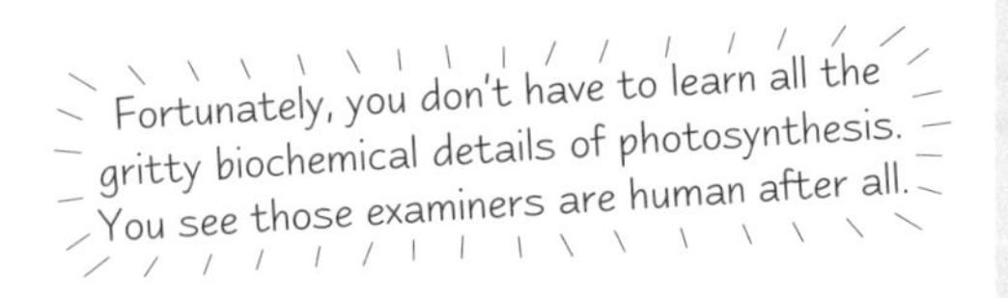

Physiologically, maize is adapted by having a **special method of photosynthesis** called **C4**. C4 photosynthesis works well at **high temperatures**. Also, it uses a **different enzyme** to **fix** carbon dioxide, which has a **high affinity for CO_2**, meaning it works well at low CO_2 concentrations. The stomata don't need to be open for long in C4 plants, which helps to **prevent water loss**.

Structurally, it's adapted to the hot environment by being able to roll its leaves up to reduce the number of stomata exposed to the atmosphere. This reduces water loss.

Sorghum Grows Well in Hot, Dry Conditions

Sorghum has many **xerophytic** characteristics — it can live in hot, dry conditions that other plants can't survive in. Xerophytes have adaptations that allow them to conserve water. **Structurally**, sorghum is very well adapted:

1) **Long, extensive roots** reach **deep into soil** to find water.
2) **Thick waxy cuticles** reduce water loss.
3) There are **relatively few stomata**, which helps **reduce water loss** from leaves.
4) The **stomata are sunken**, which means water vapour builds up round the opening, reducing the diffusion gradient and slowing diffusion.
5) Like maize, sorghum can roll its leaves.
6) Sorghum can also use **C4 photosynthesis**.

Physiologically, sorghum is adapted by containing enzymes that aren't denatured by extreme heat — so both adult and embryo plants can survive in the heat, even when they are still developing structurally.

Controlling the Abiotic Environment

*Increasing the **Rate** of **Photosynthesis** Increases **Crop Yield***

Humans can control the **abiotic** (non-living) factors that affect photosynthesis — **light intensity**, **temperature**, **water**, **CO_2**, **soil fertility** and **pH**. They encourage growth by **maximising** the **rate of photosynthesis**:

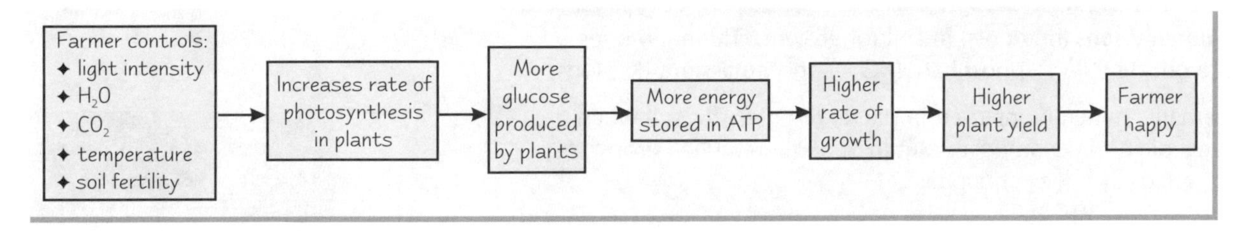

Farmers** can Create Controlled, **Artificial Environments

Farmers use large, commercial **greenhouses** to monitor and control the environment.
They aim to maintain **optimum growing conditions**:

1) Increasing the light intensity raises the photosynthesis rate — so farmers supply **artificial light** for **longer hours** than natural sunlight. But too much light can damage chloroplasts, so they have to be careful.
2) Around 25°C is the optimum temperature for photosynthesis so farmers use **paraffin heaters** in winter and **shade** and **ventilation** in summer to maintain this temperature in greenhouses.
3) A good supply of **carbon dioxide** is needed for photosynthesis. Farmers give greenhouse plants a good supply of carbon dioxide, but don't usually add more than around **0.4%**, otherwise the stomata would close up.

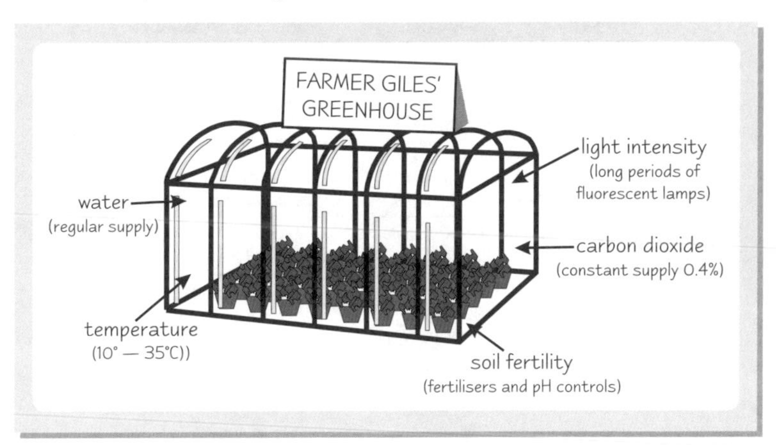

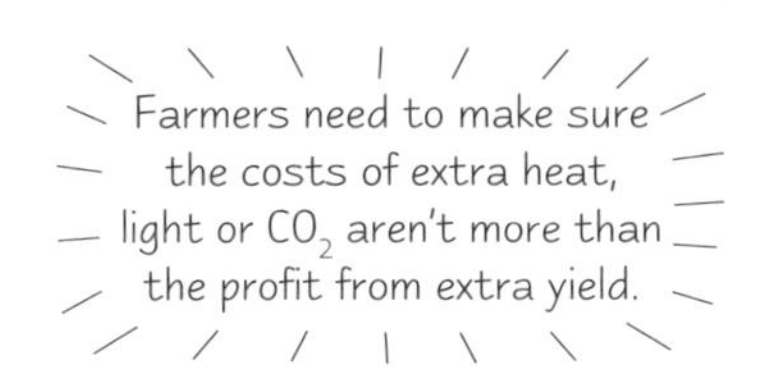

Practice Questions

Q1 Describe two ways that rice is adapted to growing in swampy conditions.

Q2 List four structural adaptations of sorghum.

Q3 How does maize reduce its water loss?

Q4 Which abiotic factors do farmers control in commercial greenhouses?

Exam Questions

Q1 Explain how a tolerance for ethanol has helped rice adapt to waterlogged soil. [4 marks]

Q2 Describe how farmers can use greenhouses to improve crop yields. [4 marks]

What an a-maize-ing topic...

Credit where credit's due — I'd like to see how well you'd survive standing in a puddle of water your entire life or living in the Sahara desert. But even with all the plants' clever adaptations, that doesn't stop humans stepping in to make them grow even better. But then, I can't quite imagine a life without bread, so I'm not complaining.

Fertilisers

Soil, fertilisers and pesticides — sounds like a recipe for maximum crop yield if ever I heard one. Learn what farmers do to make their crops grow well — and don't try this recipe at home, kids.

Fertilisers Replace Nutrients Lost from the Soil

Plants use **mineral ions** (nutrients) in soil for growth. The main ones they use are **nitrate** (NO_3^-), **phosphate** (PO_4^{3-}) and **potassium** (K^+) ions.

1) Normally, mineral ions used in plants **re-enter the soil** when the plants die and decompose. But in agriculture, when crops are harvested, this doesn't happen.
2) When nutrient levels in soil become **too low**, crops won't grow as well, which means a lower yield and smaller profit for farmers.
3) So farmers use **fertilisers** to replace the lost nutrients.

Fertilisers can be Organic or Inorganic

There are two types of fertiliser — organic and inorganic. Both have advantages and disadvantages.

1) **Organic fertilisers** (e.g. manure) are **natural products** that are put onto the soil. They decompose, releasing minerals into the soil.
2) **Inorganic fertilisers** are commercially manufactured from **mineral ions** and sprayed onto crops.

TYPE OF FERTILISER	ADVANTAGES	DISADVANTAGES
Organic	• Cheap. • Adds humus to soil, which improves soil structure. • Helps soil's water retention, so there's less leaching. • Less eutrophication. • Recycles animal waste.	• Hard to obtain in large quantities. • Expensive to transport. • Has a haphazard mineral content. • Slow release of nutrients. • Difficult to spread accurately.
Inorganic	• Can be manufactured to match the needs of a specific soil or crop. • Can be sprayed easily and accurately. • Fast release of mineral ions.	• Expensive. • Doesn't have the organic matter to 'glue' soil particles together. Repeated use can damage the soil structure. • Soil is more prone to leaching. • Causes eutrophication.

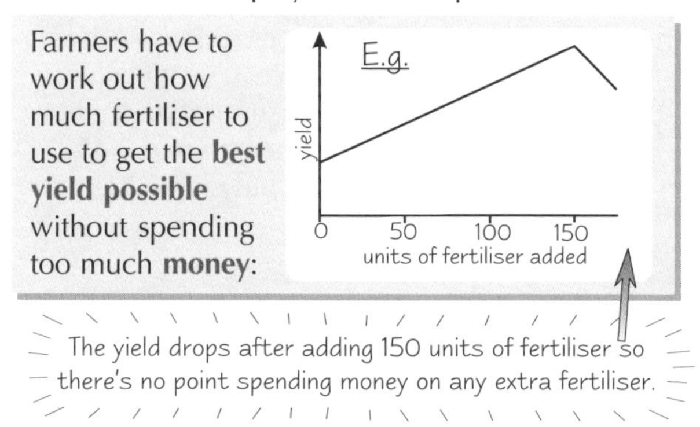

The yield drops after adding 150 units of fertiliser so there's no point spending money on any extra fertiliser.

Fertilisers can Lead to Leaching and Eutrophication

Over use of NPK (nitrate, phosphate and potassium) fertilisers can be bad news for nearby lakes, rivers and streams. Learn these flow charts to see how leaching and eutrophication happen.

1) **Leaching** is the 'washing out' of mineral ions from the soil:

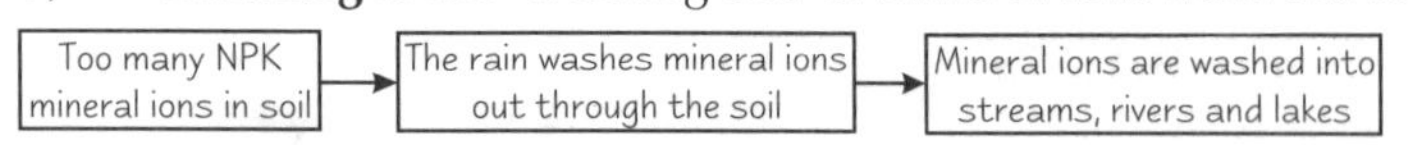

2) **Eutrophication** is what happens when too many mineral ions are leached into streams, rivers and lakes:

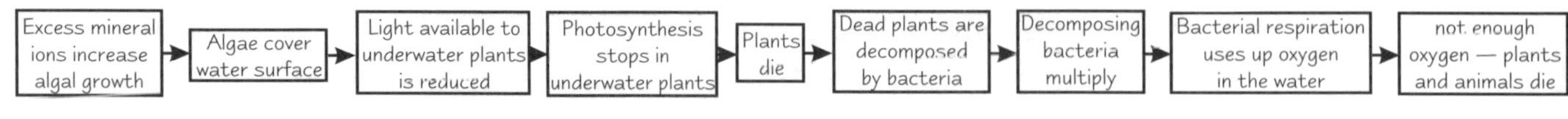

Crop Yield can be Reduced by Competition and Insects

Plants compete for light, space, nutrients, water, oxygen and carbon dioxide.

1) **Intraspecific competition** is competition between plants of the **same species**. Farmers reduce this by making sure there's enough of everything to go around. E.g. by irrigating crops and using fertilisers.
2) **Interspecific competition** is competition between **different species**, like weeds and crop plants. Farmers reduce this by using **herbicides** to **kill weeds**.

Insects can reduce crop yield directly by eating the parts of crops that humans use and indirectly by damaging photosynthetic tissue.

Pesticides

Pesticides are Used to Kill Organisms That Reduce Crop Yield

Pesticides are **manufactured chemicals** that are used to kill pests. They're split into different types, depending on what they're killing:

Type of Pesticide	Type of Organism Killed
Herbicide	Weeds
Insecticide	Insects
Fungicide	Fungi

Pesticides are **easy** and **quick** to apply. But, **resistance** builds up in the target pest, so they stop being effective after a while. Also, pesticides can have damaging effects on the **environment**:

1) Pesticides can be toxic to **non-pest organisms** as well.
2) Some pesticides (like the insecticide, **DDT**) aren't broken down by organisms — instead they build up in the organism's body. Anything that eats the organism eats the pesticide too — so toxins can therefore **enter food chains** and pass along them.
3) This leads to **bioaccumulation** — higher levels of toxin accumulate at each trophic level of a food chain.
4) **Predators** at the top of the food chain can be **killed** by the toxins from pesticides.

Biological Agents are More Environmentally Friendly

Biological agents are **organisms** used to control pests. E.g. farmers can release predators to **eat the pests** or release **bacteria** that cause **disease** in the pests. Biological control is **better for the environment** but it has some problems:

1) It's **slow**.
2) Biological agents don't eliminate all the pests, so you get **resurgence** — the pests that aren't killed reproduce and the pest population increases again.
3) Introducing new organisms into existing ecosystems can **disrupt food chains and webs**.

Integrated Systems are the Best of Both Worlds

More farmers are now using **integrated systems of pest management** — they control pests with a combination of:

1) **biological agents**;
2) **chemical pesticides**;
3) **good crop management** (e.g. timing crops so they miss seasons when pests are more common);
4) using genetically-modified **pest-resistant crops**.

Having **more than one** control method reduces the risk of pests becoming **resistant** to control. Also, it's **better for the environment** — fewer pesticide toxins and biological agents are used, so existing organism populations **aren't disrupted** as much.

Practice Questions

Q1 What does NPK stand for?

Q2 Draw up a table of advantages and disadvantages of using inorganic and organic fertilisers.

Q3 What is bioaccumulation?

Q4 What is integrated pest management?

Exam Questions

Q1 Why is a root crop yield reduced when insects eat the leaves of the crop? [4 marks]

Q2 Give two advantages and two disadvantages of using biological control instead of pesticides. [4 marks]

Farmers like to watch TOTC — Top of the Crops...

Make sure you get eutrophication sorted in your head — it's a bit tricky and can throw you if you're not careful. Remember that it's not the algae that use up all the oxygen — it's the decomposing bacteria. So next time you see someone being a bit over-keen with their fertilisers, tell them on behalf of all the little fishies that it's not big and it's not clever. Then run away.

Reproduction and its Hormonal Control

There's a lot covered on this page — the development of follicles, the menstrual cycle, implantation of the fertilised ovum... Normally all that stuff takes at least nine months, and doesn't include farm animals either.

Ova Develop Inside the Ovary Ready for Ovulation

(Blimey — that's quite a few 'ov's.)

Female foetuses begin to make eggs while they're still in the uterus. Odd, I say. Odd.
When a girl is born, she already has millions of **primary follicles**, but only some of them develop fully.

1) At puberty, sex hormones like **FSH** stimulate the primary follicles to divide by meiosis to produce **secondary oocytes** (these are the **ova** that you learn about at GCSE).
2) Each month, one of these secondary oocytes develops inside a **Graafian (mature) follicle**. The follicle travels to the surface of the ovary and bursts, which releases the secondary oocyte into the **fallopian tube**, ready for **fertilisation**. This release is called **ovulation**.

Diagram Showing the Development of Ova in the Ovary

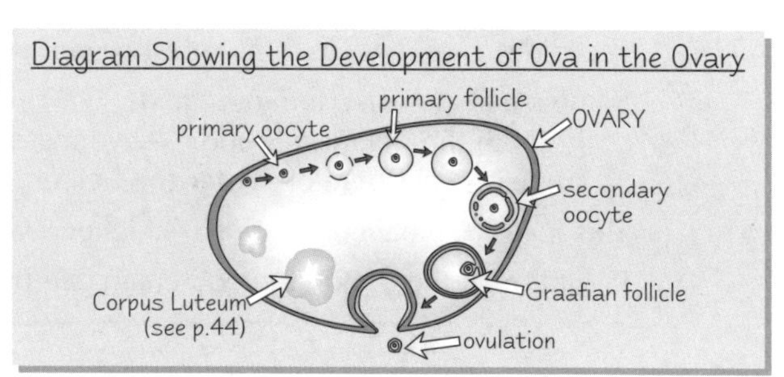

The Human Menstrual Cycle is Controlled by Hormones

The human menstrual cycle lasts about **28 days**. It involves development of a follicle in the ovary, release of an ovum, and **thickening** of the **uterus lining** so a fertilised ovum can **implant**. If there's no fertilisation, the lining breaks down and exits through the vagina. This is **menstruation**, which marks the end of one cycle and the start of another.

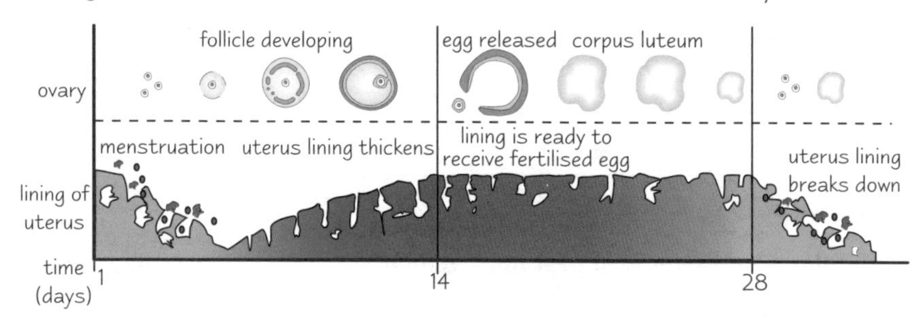

The menstrual cycle is controlled by four hormones — **Follicle-Stimulating Hormone** (FSH), **Luteinising Hormone** (LH), **oestrogen** and **progesterone**. They are either produced in the pituitary gland or in the ovaries:

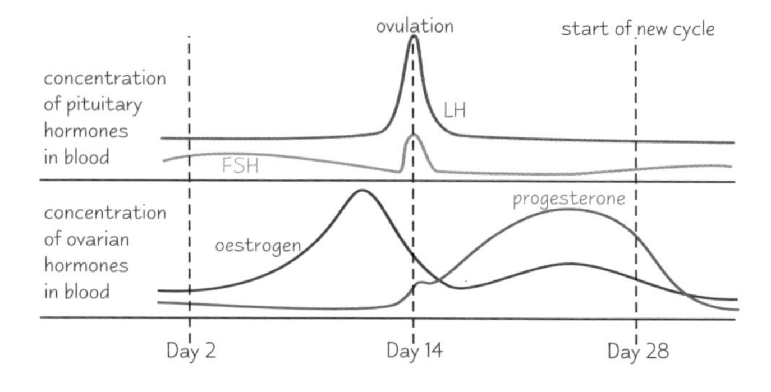

Hormones released by the pituitary —

1) **FSH** is released into the bloodstream at the **start** of the cycle and carried to the ovaries. It stimulates the development of one or more follicles, which stimulates the secretion of oestrogen.
2) **LH** is released into the bloodstream around **day 12**. It causes release of the ovum (**ovulation**). When the egg bursts out it leaves its primary follicle behind. LH helps the follicle turn into a **corpus luteum,** which is needed later.

Hormones released by the ovaries —

1) **Oestrogen** is produced by the **developing follicle**. It causes the **lining** of the uterus to **thicken**. It also **inhibits** release of FSH. This stops any more follicles maturing. A peak in oestrogen production starts a surge in FSH and LH production, which causes ovulation.
2) **Progesterone** is released by the **corpus luteum** after ovulation. Progesterone keeps the lining thick, ready for implantation if fertilisation occurs. It also inhibits release of FSH and LH. If no embryo implants, the corpus luteum dies, so progesterone production stops and FSH inhibition stops. This means the cycle starts again, with development of a new follicle.

The ovum is released around day 14 of the menstrual cycle, and must be fertilised within 24 hours. If sexual intercourse leads to fertilisation, the fertilised egg moves to the uterus where it implants in the wall. This takes up to 3 days.

Manipulation and Control of Reproduction

There are Different Treatments for Infertility

Some men and women are infertile, which prevents conception.
There are different reasons for infertility, and each reason is treated differently:

1) Some women **can't ovulate** due to **low levels of FSH** (or high levels of **oestrogen**, which **inhibits FSH**). They can be given **injections of FSH** or take a drug, **clomiphene**, to inhibit oestrogen production.
2) Some women have **blocked fallopian tubes**, which can either prevent fertilisation or stop the fertilised embryo reaching the uterus. They can be given **IVF treatment**, where their eggs are removed and fertilised with sperm ***in-vitro*** (outside their body). Then the **fertilised embryos** are **transplanted** into their uterus.
3) Some **men** have a **low sperm count**. They can be treated with **male hormones**, like **testosterone**.

Hormones Can be Used to Control Reproduction

Hormones can be used in loads of different ways to affect reproduction of humans and animals. Learn these examples.

1) **Reproduction in humans** — **Contraceptives** help women prevent unwanted pregnancies. The **combined pill** contains **synthetic** versions of **oestrogen** and **progesterone**. It **inhibits** secretion of **FSH**. No follicles can mature, so no ovulation takes place.
2) **Reproduction in farm animals** — When an animal is **in oestrus**, this means it's **fertile** and ready to mate. Animals usually display certain types of **behaviour** when they're in oestrus. For example **female cows** in oestrus become **restless** and try to **mount other cows**. This is useful for farmers, because it helps them know when to **successfully mate** their animals.
3) **Selective breeding in cattle** — Cows with **desirable features** are given an injection of the hormone **prostaglandin**, to bring them into oestrus. They are then injected with **FSH and LH** to stimulate a large number of follicles to mature in their ovaries. After ovulation, these ova can be **fertilised *in-vitro*** and the embryos transplanted into **surrogate mothers**.
4) **Synchronising breeding in sheep** — A **progesterone-containing coil** can be inserted into a sheep. When the coil is removed, the pituitary begins to produce **FSH** and ovulation happens soon after. This is used to **synchronise** breeding of sheep, which saves on costs.
5) **Increasing milk yield** — A hormone called **bovine somatotrophin** can be injected into cows to increase their milk yield.

Some people have moral and ethical issues with using hormones to mess with natural reproduction.

Practice Questions

Q1 What is "ovulation"?

Q2 Which hormones in the menstrual cycle are released by the pituitary gland?

Q3 What treatment would a woman with blocked fallopian tubes get?

Q4 Give two uses of hormones in manipulating reproduction in animals.

This book ain't over 'til the fat pig sings.

Exam Question

Q1 The combined pill is an oral contraceptive.

a) Which hormones are present in this pill? [2 marks]

b) Explain how these hormones prevent pregnancy. [3 marks]

They used to call it 'the curse' — now I know why...

Lots of hormones. You need to know their names, where they come from, what they do... where their wives go shopping. Blimey it's the end of the book. A sad, sad day for biology students everywhere. But hey, like the singing pig, you've survived all the AS biology monsters. So let's all join together for a big finale sing-song. Or maybe not. I think I'd better go now.

Answers

Section 1 — Cells
Pages 2-3 — Cells and Microscopy

1 Maximum of 6 marks available.
Advantages: Greater resolution **[1 mark]**.
Higher maximum magnification **[1 mark]**.
Disadvantages: Electron microscopes can't be used to study living tissues / processing kills living cells **[1 mark]**.
Natural colours can't be seen **[1 mark]**. They aren't portable **[1 mark]**. They are expensive **[1 mark]**.

Pages 4-5 — Functions of Organelles / Cell Fractionation

1 Maximum of 9 marks available.
You could have written about any three of the organelles given below.
Mitochondria [1 mark] — Large numbers of mitochondria would indicate that the cell used a lot of energy **[1 mark]**, because mitochondria are the site of (aerobic) respiration, which releases energy **[1 mark]**.
Chloroplasts [1 mark] — Large numbers of chloroplasts would be seen in cells that are involved in photosynthesis **[1 mark]** because the chloroplasts contain chlorophyll, which absorbs light for photosynthesis **[1 mark]**.
Ribosomes [1 mark] —
You find lots of ribosomes in cells that produce a lot of protein **[1 mark]** because ribosomes are the site where proteins are made **[1 mark]**.
Rough endoplasmic reticulum [1 mark] —
You find a lot of RER in cells that produce a lot of protein **[1 mark]** because the RER transports protein made in the attached ribosomes **[1 mark]**
Lysosomes [1 mark] — Found in cells that are old / destroy other cells **[1 mark]** because lysosomes contain digestive enzymes that can break down cells **[1 mark]**.
There are 9 marks for this question and 3 organelles have to be mentioned — so it's logical that each organelle provides 3 marks. You get a mark for mentioning the correct organelles, so you need to give 2 pieces of relevant information for each one.

2 a) Maximum of 2 marks available.
i) mitochondrion **[1 mark]**
ii) Golgi apparatus **[1 mark]**
b) Maximum of 2 marks available:
The function of the mitochondrion is to be the site of (aerobic) respiration / provide energy **[1 mark]**.
The function of the Golgi apparatus is to package materials made in the cell / to make lysosomes **[1 mark]**.
The question doesn't ask you to give the reasons why you identified the organelles as you did, so don't waste time writing your reasons down.

3 Maximum of 5 marks available.
The cells are homogenised **[1 mark]** in ice cold isotonic buffer solution **[1 mark]**. The cold makes sure that any protein digesting enzymes that are released don't digest the organelles **[1 mark]**. The buffer keeps the pH constant **[1 mark]** and the isotonic solution ensures that osmosis does not occur as this could harm the organelles **[1 mark]**.

Pages 6-7 — Plasma Membranes / Transport Across Membranes

1 Maximum of 2 marks available.
Phospholipids are arranged in a double layer / bilayer with fatty acid tails on the inside **[1 mark]**.
Fatty acid tails are hydrophobic / non-polar so they prevent the passage of water soluble molecules through the cell membrane **[1 mark]**.
Occasionally a question may ask you to show how a single layer of phospholipid molecules would arrange themselves on the surface of a container of water. You should draw the molecules with their hydrophilic phosphate heads in the water and their hydrophobic fatty acid tails sticking up into the air.

2 Maximum of 3 marks available.

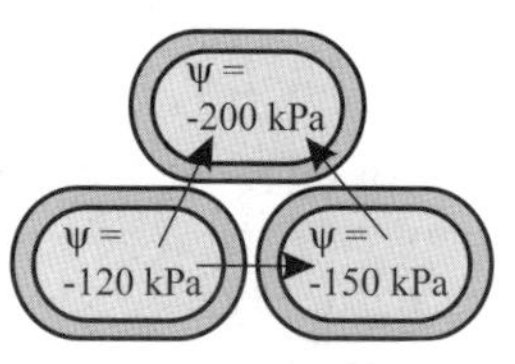

[1 mark for each correctly drawn arrow.]

Pages 8-9 — Transport Across Membranes

1 a) Maximum of 2 marks available.
Osmosis is the diffusion / movement of water molecules through a partially permeable membrane **[1 mark]** from an area of higher water potential / higher concentration of water molecules to an area of lower water potential / lower concentration of water molecules **[1 mark]**.
b) Maximum of 3 marks available.
When water moves into a plant cell by osmosis, it makes the plant cell expand and push against the cell wall **[1 mark]**, which makes the cell turgid **[1 mark]**. Turgid cells are needed to keep the plant upright **[1 mark]**.
You need to use the word 'turgid' here to get one of the marks — describing the effects of turgidity (i.e. the cell is swollen) isn't enough.

2 a) Maximum of 3 marks available.
Root hair cell has a more negative / lower water potential than the soil water **[1 mark]** because it has a greater concentration of dissolved solutes **[1 mark]**.
So water moves into the cell by osmosis from a higher water potential to a lower water potential **[1 mark]**.
b) Maximum of 4 marks available.
Root hair cells use active transport to take up minerals from the soil against a concentration gradient **[1 mark]**.
The ions attach to specific carrier proteins / pumps in the root hair cell membrane **[1 mark]**.
Molecules of ATP / adenosine triphosphate **[1 mark]** provide the energy to change the shape of the protein and move the molecules across the membrane **[1 mark]**.

Section 2 — Molecules
Pages 10-11 — Carbohydrates

1 Maximum of 7 marks available.
Glycosidic bonds are formed by condensation reactions **[1 mark]** and broken by hydrolysis reactions **[1 mark]**.
When a glycosidic bond is formed in a condensation reaction, a hydrogen **[1 mark]** from one monosaccharide combines with a hydroxyl / OH group **[1 mark]** from the other to form a molecule of water **[1 mark]**.
A hydrolysis reaction is the reverse of this **[1 mark]**, with a molecule of water being used up to split the monosaccharide molecules apart **[1 mark]**.
The last 5 marks for this question could be obtained by a diagram showing the reaction, using structural formulae.

Answers

2 *Maximum of 10 marks available.*
*Glycogen is a chain of alpha glucose molecules **[1 mark]** whereas cellulose is a chain of beta glucose molecules **[1 mark]**.*
*Glycogen's chain is compact and very branched **[1 mark]** whereas cellulose's chain is long, straight and unbranched **[1 mark]** and these chains are bonded together to form strong fibres **[1 mark]**.*
*Glycogen's structure makes it a good food store in animals **[1 mark]**. The branches allow enzymes to access the glycosidic bonds to break the food store down quickly **[1 mark]**.*
*Cellulose's structure makes it a good supporting structure in cell walls **[1 mark]**. The fibres provide strength **[1 mark]**.*
*Also, the enzymes that break glycosidic bonds can't reach the glycosidic bonds in cellulose, so they can't break cellulose down **[1 mark]**.*
In questions worth lots of marks make sure you include enough details. This question is worth 10 marks so you should include at least 10 relevant points to score full marks. Also, the question asks you to compare and contrast, so make sure you don't just describe glycogen and cellulose totally separately from each other. You need to highlight how they differ from each other, and what this means for their functions.

Pages 12-13 — Proteins

1 *Maximum of 5 marks available.*
*Proteins are made from chains of amino acids / polypeptide chains **[1 mark]**.*
*The sequence of amino acids in the chain is the protein's primary structure **[1 mark]**.*
*The amino acid chain / polypeptide coils in a certain way, which is the protein's secondary structure **[1 mark]**.*
*The coiled chain is itself folded into a specific shape, which is the protein's tertiary structure **[1 mark]**.*
*The quaternary structure is how different polypeptide chains are joined together in a protein molecule **[1 mark]**.*
The question specifically states that you don't need to describe the chemical nature of the bonds in a protein. So, even if you name them, you don't have to go into chemical details of how they're formed — no credit will be given.

2 *Maximum of 6 marks available.*
*Insulin is a globular protein **[1 mark]**.*
For this mark, mentioning that the molecule is globular is essential.
*It's a hormone that reduces blood glucose levels **[1 mark]**.*
*Its globular structure makes it soluble **[1 mark]**, so it's easily transported in the blood **[1 mark]**.*
*Disulphide bonds hold its shape together **[1 mark]** so it's small and compact **[1 mark]**.*
*Its small size means it's easily absorbed by cells **[1 mark]**.*
7 marks are listed, but the mark is given out of 6. This is common in longer exam questions. You would have to be a bit of a mind-reader to hit every mark the examiner thinks of, so to make it fair, there are more mark points than marks. You can only count a maximum of 6, though.

Pages 14-15 — Lipids

1 *Maximum of 8 marks available.*
*Triglycerides are made from a glycerol molecule **[1 mark]** and three molecules of fatty acids **[1 mark]**. They are formed by condensation reactions **[1 mark]**. These reactions result in the formation of ester bonds **[1 mark]** between the fatty acid and glycerol molecules, with the production of a molecule of water for each fatty acid added **[1 mark]**.*
*Triglycerides are broken up by hydrolysis reactions **[1 mark]**, which are the reverse of condensation reactions **[1 mark]**, with one molecule of water being added for each fatty acid that's released **[1 mark]**.*
It would be possible to get all the marks in this question by using labelled diagrams, as long as all the points listed have been illustrated.

2 *Maximum of 8 marks available.*
*A triglyceride consists of glycerol **[1 mark]** and three fatty acid molecules **[1 mark]**. A phospholipid has the same basic structure, but one of the fatty acids is replaced by a phosphate group **[1 mark]**.*
*Triglycerides are hydrophobic / repel water **[1 mark]**. This is a property of the hydrocarbon chains that are part of the fatty acid molecules **[1 mark]**. The phosphate group in a phospholipid is hydrophilic / attracts water **[1 mark]**, because it's ionised **[1 mark]**. This means that the phospholipid has a hydrophilic 'head' and a hydrophobic 'tail' **[1 mark]**.*
The words 'head' and 'tail' are not essential, as long as you have got across the idea that the molecule is partly hydrophobic and partly hydrophilic.

Pages 16-17 — Biochemical Tests for Molecules / Chromatography

1 *Maximum of 14 marks available.*
*Test for starch **[1 mark]** by adding iodine in potassium iodide **[1 mark]**. A positive reaction would have a blue-black colour **[1 mark]**.*
*Test for reducing sugar **[1 mark]** by heating with Benedict's reagent **[1 mark]**. The formation of a brick red precipitate would indicate the presence of reducing sugar **[1 mark]**.*
Instead of mentioning red precipitate, you could just describe a colour change from blue to brick red.
*Test for non-reducing sugars **[1 mark]** by boiling with hydrochloric acid and then neutralising before doing the Benedict's test **[1 mark]**.*
*Test for proteins **[1 mark]** by adding sodium hydroxide solution, then copper (II) sulphate solution **[1 mark]**. Protein is indicated by a purple colour **[1 mark]**.*
*Test for lipids **[1 mark]** by adding ethanol then mixing with water **[1 mark]**. Lipid is indicated by a milky colour **[1 mark]**.*
For full marks, make sure you've mentioned what each biochemical group is, how you test for it and what results you might get.

2 *Maximum of 7 marks available.*
*Grind up the leaves in a solvent **[1 mark]**.*
*Add a drop of the extract to strip of chromatography paper **[1 mark]**.*
*Place the end of the paper into a solvent so that the solvent rises up the paper **[1 mark]**.*
*Measure the distance that the solvent has travelled **[1 mark]** and the distance that each spot of pigment has travelled **[1 mark]**.*
*To identify the pigments calculate their R_f values **[1 mark]** by using the formula R_f = distance travelled by pigment/distance travelled by solvent **[1 mark]**.*
Sometimes it is easier or quicker to describe something in a diagram rather than in words (e.g. calculating R_f values). You can get full marks this way as long as the diagram shows what the examiner wants to know.

Pages 18-19 — Action of Enzymes

1 *Maximum of 3 marks available.*
*Enzymes speed up / catalyse all biological reactions **[1 mark]** without being used in the reaction themselves **[1 mark]**.*
*They do this by lowering the activation energy that's required before the reaction can start **[1 mark]**.*

2 *Maximum of 8 marks available.*
'Lock and key' model —
*The enzyme and the substrate have to fit together at the active site of the enzyme **[1 mark]**.*
*This creates an enzyme-substrate complex **[1 mark]**.*
*The active site then causes changes in the substrate **[1 mark]**.*
This mark could also be gained by explaining the change (e.g. bringing molecules closer together, or putting a strain on bonds).
*The change results in the substrate being assembled / broken down **[1 mark]**.*
*'Induced fit' model — has the same basic mechanism as the lock and key model **[1 mark]**.*
*The difference is that the substrate is thought to cause a change in the enzyme's active site **[1 mark]**, which enables a better fit **[1 mark]**.*

Answers

Pages 20-21 — Factors that Affect Enzyme Activity

1 *Maximum of 8 marks available, from any of the 9 points below.*
If the solution is too cold, the enzyme will work very slowly ***[1 mark]****.*
This is because, at low temperatures, the molecules move slowly and collisions are less likely between enzyme and substrate molecules ***[1 mark]****.*
The marks above could also be obtained by giving the reverse argument — a higher temperature is best to use because the molecules will move fast enough to give a reasonable chance of collisions.
If the temperature gets too high, the reaction will stop ***[1 mark]****.*
This is because the enzyme is denatured ***[1 mark]*** *— the active site changes and will no longer fit the substrate* ***[1 mark]****.*
Denaturation is caused by increased vibration breaking bonds in the enzyme ***[1 mark]****.*
Enzymes have an optimum pH ***[1 mark]****.*
pH values too far from the optimum cause denaturation ***[1 mark]****.*
Explanation of denaturation here will get a mark only if it hasn't been explained earlier.
Denaturation by pH is caused by disruption of ionic bonds, which destabilises the enzyme's tertiary structure ***[1 mark]****.*

2 *Maximum of 4 marks available.*
Chemical X is an enzyme inhibitor ***[1 mark]***
Reason — it reduces an enzyme controlled reaction ***[1 mark]***
The inhibitor is probably competitive ***[1 mark]***
Reason — increasing the concentration of the inhibitor makes it more effective, because if there are a lot of inhibitor molecules they're more likely to reach active sites before the substrate molecules and will block them ***[1 mark]****.*

Section 3 — Systems

Pages 22-23 — Tissues / Surface Area to Volume Ratio

1 *Maximum of 3 marks available.*
Red blood cells don't have a nucleus, so there's more space for carrying oxygen ***[1 mark]****.*
They have a bi-concave shape to increase their surface area for absorbing oxygen ***[1 mark]****.*
They have an elastic membrane so that they can squeeze through the narrow capillaries and take oxygen to cells ***[1 mark]****.*

2 *Maximum of 4 marks available.*
Humans are large multi cellular organisms ***[1 mark]****.*
The surface area : volume ratio is small in large organisms ***[1 mark]****, which makes diffusion too slow* ***[1 mark]****.*
Humans need specialised organs for gaseous exchange with a large enough surface area to keep all their cells supplied with enough oxygen and to remove CO_2 ***[1 mark]****.*

Pages 24-25 — Organs and Blood Transport / Tissue Fluid

1 *Maximum of 4 marks available, taken from any of the points below.*
They have thick, muscular walls ***[1 mark]*** *to cope with the high pressure caused by the heartbeat* ***[1 mark]****.*
They have elastic tissue in the walls ***[1 mark]*** *so they can expand to cope with the high pressure caused by the heartbeat* ***[1 mark]****.*
The inner lining (endothelium) is folded ***[1 mark]*** *so that the artery can expand when the heartbeat causes a surge of blood* ***[1 mark]****.*

2 *Maximum of 5 marks available.*
Tissue fluid moves out of capillaries due to a pressure gradient ***[1 mark]****.*
At the arteriole end, pressure in capillary beds is greater than pressure in tissue fluid outside capillaries ***[1 mark]****.*
This means fluid from blood is forced out of the capillaries ***[1 mark]****.*
Fluid loss means the water potential of blood capillaries is lower than that of tissue fluid ***[1 mark]****.*
So fluid moves into the capillaries at the venule end by osmosis ***[1 mark]****.*

Pages 26-27 — Lungs and Ventilation

1 *Maximum of 4 marks available.*
Fick's law states that the rate of diffusion is affected by surface area, difference in concentration / concentration gradients and the thickness of the exchange surface / the length of the diffusion pathway ***[1 mark]****.*
For this mark you need to mention the 3 factors affecting diffusion rate, according to Fick's law.
There are lots of alveoli, which gives a large surface area for gas exchange ***[1 mark]****.*
The alveolar epithelium is only one cell thick, so there's a short diffusion pathway ***[1 mark]****.*
They also maintain steep concentration gradients of the repiratory gases to speed up the diffusion rate ***[1 mark]****.*
Your answer must only describe alveolar adaptations that are relevant to the factors in Fick's law. Details of how concentration gradients are maintained in the alveoli aren't required here.

Pages 28-29 — The Heart and the Cardiac Cycle

1 *Maximum of 3 marks available.*
Pressure increases in the atria when they contract and in the ventricles when they contract ***[1 mark]****.*
Pressure decreases in the atria when they relax and in the ventricles when they relax ***[1 mark]****.*
There is always more pressure in the left ventricle than the right ventricle, because of the thicker muscle walls producing more force ***[1 mark]****.*
This question doesn't ask you to describe the cardiac cycle — it specifically asks you to describe the pressure changes during contraction and relaxation. Make sure you mention both atria and ventricles in your answer.

2 *Maximum of 6 marks available.*
The valves only open one way ***[1 mark]****.*
Whether they open or close depends on the relative pressure of the heart chambers ***[1 mark]****.*
If the pressure is greater behind a valve (i.e. there's blood in the chamber behind it) ***[1 mark]****, it's forced open, to let the blood travel in the right direction* ***[1 mark]****.*
When the blood goes through the valve, the pressure is greater above the valve ***[1 mark]****, which forces it shut, preventing blood from flowing back into the chamber* ***[1 mark]****.*
Here you need to explain how valves function in relation to blood flow, rather than just in relation to relative pressures.

Pages 30-31 — Effects of Exercise

1 *Maximum of 4 marks available.*
Blood flow increases in skeletal muscle and heart because muscles require more oxygen during aerobic respiration ***[1 mark]****.*
Blood flow decreases in the digestive system because demand for nutrients is lower than demand for oxygen / so more blood is available for the skeletal muscles ***[1 mark]****.*
It remains the same in the brain and kidneys because the brain needs a constant supply of oxygen ***[1 mark]****, and to prevent a build-up of toxins in blood as it travels through the kidneys* ***[1 mark]****.*
As this question shows, it's not always 1 mark for each point you make. In this case you have to say what happens and why to get just one mark. So don't make 4 points and decide you've done enough — make sure you've actually answered the question.

2 *Maximum of 4 marks available.*
The low blood pressure would be detected by pressure receptors in the arteries ***[1 mark]****.*
The pressure receptors would send impulses to the cardiovascular centre in the medulla in the brain ***[1 mark]****.*
The cardiovascular centre would then send impulses to the SAN to speed up the heart rate ***[1 mark]****.*
This would cause blood pressure to increase ***[1 mark]****.*
Make sure you've said that speeding up the heart rate is done to increase blood pressure. It might seem a bit obvious, but you should still write it to show you understand the whole process.

Answers

3 *Maximum of 5 marks available.*
*Pulmonary ventilation would increase **[1 mark]** because breathing rate would increase **[1 mark]** and so would tidal volume **[1 mark]**. This is because of the increased demand for oxygen for aerobic respiration **[1 mark]** and decrease in pH due to extra CO_2 in the blood **[1 mark]**.*

Section 4 — Making Use of Biology

Pages 32-33 — Isolation of Enzymes

1 a) *Maximum of 3 marks available.*
Any 3 of these:
*They're easier to isolate than intracellular enzymes **[1 mark]**.*
*You don't need to break open cells to get to them **[1 mark]**.*
*You don't need to separate them from all the other stuff inside a cell — they're usually secreted from the cell on their own **[1 mark]**.*
*They're more stable than intracellular enzymes, which are often only stable within the cell environment **[1 mark]**.*

b) *Maximum of 2 marks available.*
Any 2 of these:
*May compete for nutrients **[1 mark]**.*
*May produce toxic substances **[1 mark]**.*
*Production time, and therefore money, wasted cleaning the equipment **[1 mark]**.*
Plus any other sensible answer.
This question is basically just common sense. Sometimes you'll be asked questions that require you to think around a topic for yourself a bit. They'll always be very straightforward questions like this, though.

2 a) *Maximum of 2 marks available.*
*Immobilised enzymes aren't free in solution **[1 mark]**.*
*Instead they're absorbed on / trapped in / encapsulated in an inert (non-reactive) material **[1 mark]**.*
It might seem like these two points are saying the same thing but they are actually different, so you have to make sure you mention both of them.

b) *Maximum of 2 marks available. Any 2 of these:*
*Does not contaminate the product **[1 mark]**.*
*Can be easily recovered and used again **[1 mark]**.*
*More stable **[1 mark]**.*

Pages 34-35 — Mitosis and the Cell Cycle

1 a) *Maximum of 6 marks available.*
*A = Metaphase **[1 mark]**, because the chromosomes are lining up at the equator **[1 mark]**.*
*B = Telophase **[1 mark]**, because the cytoplasm is dividing to form two new cells **[1 mark]**.*
*C = Anaphase **[1 mark]**, because the centromeres have divided and the chromatids are moving to opposite poles **[1 mark]**.*
If you've learned the diagrams of what happens at each stage of mitosis, this should be a breeze. That's why it'd be a total disaster if you lost three marks for forgetting to give reasons for your answers. Always read the question properly and do exactly what it tells you to do.

b) *Maximum of 3 marks available.*
*X = Chromatid **[1 mark]**.*
*Y = Centromere **[1 mark]**.*
*Z = Spindle fibre **[1 mark]**.*

2 a) *Maximum of 2 marks available.*
*Interphase **[1 mark]**, during the S or Synthesis stage **[1 mark]**.*

b) *Maximum of 2 marks available.*
*Mitosis **[1 mark]**, during the prophase stage **[1 mark]**.*
Sometimes you can pick up marks even if you're only half right. You might know spindle fibres are formed during mitosis, but you've forgotten which stage. It's worth mentioning mitosis anyway, and then having a guess. You won't lose marks if your guess is wrong, and even if it was you'd have picked up one mark just for saying mitosis. Better than leaving it blank.

Pages 36-37 — Meiosis

1 a) *Maximum of 3 marks available.*
*A = 46 **[1 mark]**.*
*B = 23 **[1 mark]**.*
*C = 23 **[1 mark]**.*

b) *Maximum of 2 marks available, from any of the points below.*
*Normal body cells have two copies of each chromosome, which they inherit from their parents **[1 mark]**.*
*Gametes have to have half the number of chromosomes so that when fertilisation takes place, the resulting embryo will have the correct diploid number **[1 mark]**.*
*If the gametes had a diploid number, the resulting offspring would have twice the number of chromosomes that it should have **[1 mark]**.*

2 a) *Maximum of 2 marks available.*
*A is an extrinsic protein / receptor protein **[1 mark]**.*
*B is an intrinsic protein / carrier protein **[1 mark]**.*

b) *Maximum of 2 marks available.*
*Tertiary structure gives the receptor protein a particular shape **[1 mark]***
*Only molecules with a specific shape will fit into (or bind with) the protein **[1 mark]***

Pages 38-39 — Basic Structure of DNA and RNA / Replication of DNA

1 *Maximum of 2 marks available.*
*The long length and coiled nature of DNA molecules allows the storage of vast quantities of information **[1 mark]**.*
You only get the mark here if you've mentioned the length <u>and</u> the coiled nature.
*Good at replicating itself because of the two strands being paired **(1 mark)**.*
This question is worth 2 marks, so make sure you mention at least two things.

2 *Maximum of 5 marks available.*
*Nucleotides are joined by condensation reactions **[1 mark]**.*
*This happens between the phosphate group and the sugar of the next nucleotide **[1 mark]**.*
*The DNA strands join through hydrogen bonds **[1 mark]** between the base pairs **[1 mark]**.*
*The final mark is given for at least one accurate diagram showing at least one of the above processes **[1 mark]**.*
As the question asks for a diagram make sure you do at least one.
E.g., something along the lines of this:

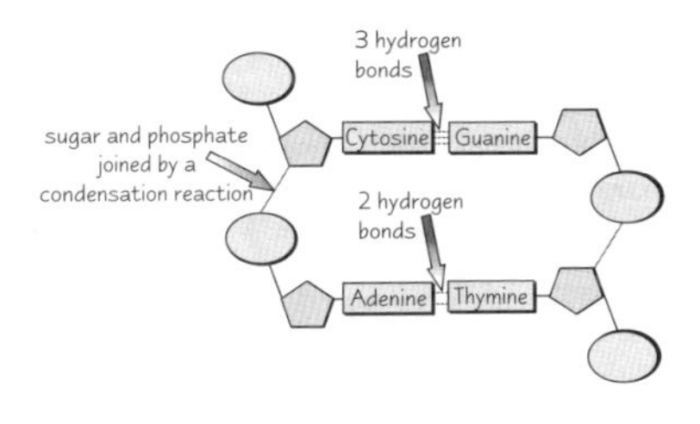

3 *Maximum of 6 marks available.*
*DNA strands uncoil and separate **[1 mark]**.*
*Individual free DNA nucleotides pair up with their complementary bases on the template strand **[1 mark]**.*
*DNA polymerase joins the individual nucleotides together **[1 mark]**.*
Students often forget to mention this enzyme in their answers.
*Hydrogen bonds then form between the bases on each strand **[1 mark]**.*
*Two identical DNA molecules are produced **[1 mark]**.*
*Each of the new molecules contains a single strand from the original DNA molecule and a single new strand **[1 mark]**.*

Answers

Pages 40-41 — The Genetic Code / Types of RNA

1 *Maximum of 2 marks available.*
*A gene is a section of DNA **[1 mark]** that codes for a particular polypeptide / protein **[1 mark]**.*

2 *Maximum of 2 marks available.*
UUA CGU CCG AGA
[2 marks if completely correct. 1 mark if mostly correct. No marks if only up to half right.]

Pages 42-43 — Protein Synthesis

1 *Maximum of 2 marks available.*
*A codon is a triplet of bases found on a mRNA molecule **[1 mark]**.*
*An anticodon is a triplet of bases found on a tRNA molecule **[1 mark]**.*
This question is only worth two marks so the examiner only expects a brief answer.

2 *Maximum of 10 marks available.*
*Transcription happens inside the nucleus and translation outside in the cytoplasm **[1 mark]**.*
*A section of DNA is uncoiled by breaking hydrogen bonds **[1 mark]**.*
*mRNA makes a copy of an uncoiled section of DNA **[1 mark]**.*
*The mRNA travels outside the nucleus to a ribosome **[1 mark]**.*
*The codons on the mRNA are paired with anticodons on a molecule of tRNA **[1 mark]**.*
*The tRNA molecules are carrying amino acids **[1 mark]** which line up and are joined by peptide bonds with an enzyme **[1 mark]**.*
You must mention that an enzyme is involved to get this mark.
*Specific codons / base triplets code for specific amino acids **[1 mark]**.*
*A polypeptide chain is formed **[1 mark]** — the primary structure of a protein **[1 mark]**.*

Pages 44-45 — Recombinant DNA

1 a) *Maximum of 2 marks available.*
*Extract all the DNA from the cell **[1 mark]**.*
*Identify the gene and use a restriction endonuclease enzyme to cut it out **[1 mark]**.*

b) *Maximum of 2 marks available.*
*Add the same restriction endonuclease enzyme to bacterial plasmids (the bacterial DNA) **[1 mark]**.*
*Insert the gene into the treated plasmids and add ligase to attach the gene to bacterial DNA **[1 mark]**.*
Don't panic if the question mentions organisms you haven't learnt about. If you read it carefully it will contain familiar ideas. This is the examiners' way of seeing whether you understand the main ideas and if you can apply what you've learnt.

2 *Maximum of 3 marks available.*
*The viral DNA is combined with the useful gene to make recombinant DNA **[1 mark]**.*
*The virus then infects a bacterium by injecting its DNA strand into it **[1 mark]**.*
*The viral DNA is then integrated into the bacterium's DNA **[1 mark]**.*
You could easily answer this question by drawing the diagram shown on p.45.

Pages 46-47 — Recombinant DNA / The Ethics of Genetic Engineering

1 a) *Maximum of 1 mark available.*
*Replica plating **[1 mark]**.*

b) *Maximum of 5 marks available.*
*Bacteria that have been transformed / have taken up the plasmid contain both the human gene and the gene for antibiotic resistance **[1 mark]**.*
*After being mixed with plasmids, the bacteria are cultured on an agar plate called the master plate **[1 mark]**.*
*Then a sterile velvet pad is pressed onto the master plate, which picks up some bacteria from each colony **[1 mark]**.*
*The pad is pressed onto a fresh agar plate, containing an antibiotic, and some of the bacteria from each colony are transferred onto the agar surface **[1 mark]**.*
*Only transformed bacteria can grow and reproduce on the replica plate — the others don't contain the antibiotic-resistant gene, so they stop growing **[1 mark]**.*

2 *Maximum of 6 marks available, taken from any of the points below, although no marks are awarded if only disadvantages or only advantages are mentioned.*
Advantages:
*It helps the development of specific medicines to treat diseases **[1 mark]**.*
*Faulty genes can be identified and replaced, to prevent genetically inherited diseases **[1 mark]**.*
*Parents can make sure babies don't have faulty genes before birth **[1 mark]**.*
*Crops can be engineered to give better yields **[1 mark]**.*
*Crops can be engineered to grow in new habitats **[1 mark]**.*
Plus any other sensible answer.
Disadvantages:
*There's a risk of foreign genes entering non-target organisms and disrupting functions **[1 mark]**.*
*Risk of accidental transfer of unwanted genes, which could damage the recipient **[1 mark]**.*
*The possible development of a 'genetic underclass' if parents who can afford it can determine the characteristics of their child **[1 mark]**.*
*Religious concerns about 'playing God' **[1 mark]**.*
*The evolutionary consequences of genetic engineering are unknown **[1 mark]**.*
*Health concerns over human consumption of genetically modified food **[1 mark]**.*
*Doctors might be put under pressure to implant embryos that could provide transplant material for siblings **[1 mark]**.*
Plus any other sensible answer.
It's important that you learn both sides of the argument, rather than just your own opinions. You throw away all 6 marks of this question if you don't mention both disadvantages and advantages.

Pages 48-49 — Immunology / Blood Groups

1 *Maximum of 6 marks available.*
*Each B-cell / B-lymphocyte has a specific antibody on its membrane **[1 mark]**.*
*When a B-cell meets a complementary foreign antigen, the antigen binds to the antibody's receptor site **[1 mark]**.*
*This stimulates the B-cell to release the antibody into the blood **[1 mark]**.*
*The B-cell then divides by mitosis, producing many clones / plasma cells **[1 mark]**.*
*The plasma cells / clones release large quantities of the same antibody, hopefully killing the pathogen **[1 mark]**.*
*Most of the plasma cells then die, but some become memory B-cells, which provide us with immunity against a pathogen if it enters the body again **[1 mark]**.*
Don't worry if your answer is organised differently to the one above (as long as you've included the same points) but do make sure it has a sensible structure.

Answers

2 *Maximum of 7 marks available.*
*People of blood group A have type A antigens / agglutinogens on their red blood cells **[1 mark]** and type B antibodies / agglutinins in their blood plasma **[1 mark]**.*
*People of blood group B have type B antigens / agglutinogens on their red blood cells **[1 mark]** and type A antibodies / agglutinins in their blood plasma **[1 mark]**.*
*If these two blood groups are mixed in one person after a transfusion, agglutination occurs **[1 mark]**. Type A antibodies / agglutinins bind to type A antigens / agglutinogens in the foreign blood, and type B antibodies / agglutinins bind to the type B antigens / agglutinogens in the foreign blood **[1 mark]**. This can be dangerous as it leads to clumping of blood cells, which can block arteries **[1 mark]**.*
You can use the terminology 'antigens and antibodies' and still get the marks. But that's no excuse not to learn the proper terminology ('agglutinogens' and 'agglutinins'), because another question might ask you specifically about these terms.

Pages 50-51 — Genetic Fingerprinting / Polymerase Chain Reaction

1 a) *Maximum of 1 mark available.*
*Genetic / DNA fingerprinting **[1 mark]**.*

b) *Maximum of 5 marks available.*
*A blood/semen/sweat/hair/skin sample would be taken from the suspect **[1 mark]**.*
*The DNA from the sample would be cut into fragments by the specific restriction endonucleases **[1 mark]**.*
*The DNA fragments from both samples would be separated out by size using electrophoresis **[1 mark]**.*
*Radioactive gene probes would be used to make the DNA fragment patterns of both samples visible **[1 mark]**.*
*The bands on the photographic film would be compared to those of DNA samples found at the scene of the crime **[1 mark]**.*
'Explain' in an exam question tests your ability to apply your knowledge. You must give your answer in the context of the question. Simply describing the technique of genetic fingerprinting wouldn't get you full marks.

Pages 52-53 — Adaptations of Cereals / Controlling the Abiotic Environment

1 *Maximum of 4 marks available.*
*Waterlogged soil has very little oxygen for respiration **[1 mark]** so rice uses anaerobic respiration **[1 mark]**. Anaerobic respiration produces lots of toxic ethanol **[1 mark]**. Rice is tolerant to ethanol and not poisoned by the excess **[1 mark]**.*
You don't need to go into loads of details about anaerobic respiration — you just need the details that are relevant to this question about rice plants. You won't get marks for any extra details you put in.

2 *Maximum of 4 marks available, from any of the answers below:*
*Farmers manipulate the environment in carefully controlled greenhouses. They increase light intensity **[1 mark]**, supply carbon dioxide at constant 0.4% **[1 mark]** and keep temperature at 25°C optimum for photosynthesis **[1 mark]**. They supply water **[1 mark]** and fertiliser **[1 mark]** and they control soil pH **[1 mark]**.*

Pages 54-55 — Fertilisers / Pesticides

1 *Maximum of 4 marks available.*
*Leaves contain photosynthetic tissue **[1 mark]**.*
*If leaves are eaten then photosynthesis is reduced **[1 mark]**.*
Lower photosynthesis means less food,
*so respiration decreases **[1 mark]**.*
*Growth slows and crop yields are lower **[1 mark]**.*
Make sure you put in every stage in the sequence of events — so you get the full 4 marks. Don't just assume that a stage is too obvious to mention.

2 *Maximum of 4 marks available, from any of the answers below (but at least one advantage and one disadvantage must be mentioned):*
*Advantages — fewer toxins absorbed by non-target organism **[1 mark]**.*
*Pests don't build up resistance to biological agents like they do to pesticides **[1 mark]**.*
*Disadvantages — resurgence of pests **[1 mark]**.*
*Release of biological agents into environment might damage food chains / webs **[1 mark]**.*
*Slow to work **[1 mark]**.*
It's good to learn some examples of biological agents and pesticides.

Pages 56-57 — Reproduction and its Hormonal Control / Manipulation and Control of Reproduction

1 a) *Maximum of 2 marks available.*
*Oestrogen **[1 mark]**.*
*Progesterone **[1 mark]**.*

b) *Maximum of 3 marks available.*
*These hormones inhibit FSH **[1 mark]**.*
*FSH stimulates development of follicles, so inhibiting it means no follicles can mature **[1 mark]**.*
*Ovulation can't take place **[1 mark]**.*

Index

A

abiotic environment 53
action of enzymes 19
activation energy 18
active transport 7, 8
adaptations of cereals 52
adenine 38
adenosine triphosphate 8
aerenchyma 52
agglutinins 49
agglutinogens 49
alleles 40
alpha-glucose 11
alveolar epithelium 26
alveoli 22, 26
amino acids 12, 40
amino group 12
amylopectin 11
amylose 11
anaphase 34
animal cells 2, 4
anthers 36
antibodies 48, 49
antigens 48, 49
aortic bodies 31
aqueous solutions 6
arteries 24
arterioles 24
artificial environments 53
asexual reproduction 34
atria 28, 29
atrioventricular node (AVN) 28
atrioventricular valves 28

B

B-lymphocytes 48
Bacillus subtilis 32
bacteria 45, 47
base triplets 40
Benedict's test 16
beta-glucose 11
bilayer 6
biochemical tests for molecules 16
biological agents 55
biological catalysts 18
biotechnology 32
biuret test 17
blood 22, 26, 28, 29, 30, 49
blood plasma 25, 49
blood pressure 30
blood transfusions 49
blood transport 24
blood vessels 24
bonds 12
breakdown reaction 18
breathing 26, 30
brush border 2
buffer solution 5

C

capillaries 22, 24, 25
capillary beds 24
capsules 2
carbohydrates 10, 11, 12
carbon 14
carbon dioxide 22, 26
carboxyl group 12
cardiac cycle 28, 29
cardiac muscles 29
cardiac output 30
carotid arteries 30
carotid bodies 31
carrier proteins 6, 8
cartilage 26
catalyst 18, 20
cell cycle 35
cell division 34, 39, 40
cell recognition 6
cell structure 2
cells 2, 3, 22
cellulases 32
cellulose 4, 10, 11
centrioles 34
channel proteins 6, 8
chemical pesticides 55
chemoreceptors 31
chloroplast 5
cholesterol molecules 6
chromatin 4
chromatography 17
chromosomes 34, 37, 40
clones 48
closed double circulation 24
collagen 13
competitive inhibitors 21
condensation reactions 10, 12, 14, 38
contraceptives 57
control of reproduction 57
controlling the abiotic environment 53
cords 28
corpus luteum 56
cytoplasm 5, 9
cytosine 38

D

degenerate 40
denatured 20
deoxygenated blood 24
deoxyribonucleic acid 38
deoxyribose sugar 38
diabetes mellitus 33
diabetics 47
differential centrifugation 5
diffusion 7
diffusion pathway 24
digestion 2, 9
digestive enzymes 4, 9
diploid 36
disaccharides 16
disulphide bridges 12
DNA 34, 36, 38, 43, 44, 46, 50

E

elastic fibres 26
electrical impulses 28
electromagnets 3
electron microscopes 2, 3, 4
electrons 3, 16
electrophoresis 50
emulsion tests 16
endocytosis 7, 9
endoplasmic reticulum 4, 9
endothelium 24, 26
enzyme activity 20, 21
enzyme concentration 20
enzyme-substrate complex 18
enzymes 9, 11, 18, 32, 43, 50
epithelial cells 2
epithelial tissue 22
ethanoic orcein 35
eukaryotic cells 2
eutrophication 54
exchange materials 23
exchange surfaces 22
exercise 30
exocytosis 7, 9
expiration 27
extracellular enzymes 32

F

facilitated diffusion 6, 7, 8
fats 14
fermenter 32
fertilisation 36
fertilisers 54
fibrous proteins 13

Index

Fick's law 7, 26
fighting diseases 22
flaccid cells 8
flat cells 22
fluid mosaic model 6
Follicle-Stimulating Hormone (FSH) 56
forensic investigations 50

G

gametes 36
gaseous exchange 26, 31
gene probes 50
genes 40, 43, 47
genetic code 40
genetic engineering 47
genetic fingerprinting 50
globular proteins 13, 18
glucose 6, 13
glucose oxidase 33
glycerol 14
glycogen 11
glycoproteins 4, 6, 48
glycosidic bonds 10
Golgi apparatus 4, 9
grana 5
granulocytes 22
guanine 38

H

haemoglobin 13, 22, 26
haemophiliacs 47
haploid 36
heart 28
heart rate 30, 31
homogenised 5
homologous pairs 40
hormonal control 56
hormones 56, 57
hydrocarbons 14
hydrogen 14
hydrogen bonds 11, 12, 38, 44
hydrogen peroxidase 33
hydrolysis 10, 12, 14
hydrophilic 6, 13, 15
hydrophobic 15

I

immobilised enzymes 33
immobilised lactase 33
immunology 48
induced fit model 19
industrial fermenters 47
infertility 57
inorganic fertilisers 54
insulin 13
intercostal muscles 26
interphase 34
intracellular enzymes 32
iodine 16
ionic bonds 12, 20
isolation of enzymes 32
isotonic 5

K

kidneys 30
kilopascals 25
kinetic energy 20

L

lamellae 5
leaching 54
LH 56
light microscopes 3
lipases 32
lipids 4, 6, 10, 12, 14, 15
liquid tissues 22
Lock and Key model 19
lungs 22, 26, 27
luteinising hormone (LH) 56
lymph 25
lymph nodes 25
lymphatic system 25
lymphocytes 22, 48
lysosomes 4, 9

M

mammalian closed circulatory system 24
marker genes 46
medulla oblongata 26, 31
meiosis 34, 36, 37
membranes 4, 6
menstrual cycle 56
metabolic pathways 43
metaphase 34
micro-organisms 32, 45
microfibrils 11
micrometre 3
microscopy 2, 3
microvilli 2
mitochondria 4, 6
mitosis 34, 35, 48
molecules 6, 10
monocytes 22
monomers 10
mononucleotides 38
monosaccharides 10, 16
mRNA 42, 44
multi-cellular organisms 22
myogenic 28

N

nanometre 3
nerve impulses 30
nerves 28
nitrate 54
non-competitive inhibitors 21
nuclear membranes 4
nucleic acids 38
nucleotides 38
nucleus 2, 4
nutrients 54

O

oestrogen 56
oils 14
organelles 2, 4, 5, 6, 22
organic fertilisers 54
organisms 23
organs 24
osmosis 5, 7, 8, 25
ovules 36
oxygen 13, 14, 22, 26

P

passive movement 7
pathogens 48
pentose sugar 38
pepsin 20
peptide bonds 12, 42
pest-resistant crops 55
pesticides 55
pH 20, 31
phage virus 45
phagocytes 48
phagocytosis 9
phenotype 18, 43
phosphate 6, 54
phospholipids 6, 15
phosphorus 50
photosynthesis 5, 52, 53
pinocytosis 9
plant cells 2, 4, 8
plasma 22, 25
plasma cells 48
plasma membranes 4, 6
plasmids 44, 47

Index

pleural membranes 26
polymerase chain reaction 50, 51
polymerase enzymes 44
polymers 38
polypeptide chain 12
polysaccharides 11
polyunsaturated lipids 14
potassium 54
predators 55
pressure receptors 30
primary response 49
primary structure of proteins 12
primers 50
progesterone 56
prokaryotic cells 2
promoter gene 47
prophase 34
proteases 32
protein production 47
protein synthesis 42, 43
proteins 4, 6, 10, 12, 13, 17, 25, 40, 41, 43,
pulmonary arteries 24, 29
pulmonary veins 24
pulmonary ventilation 30, 31

Q

quaternary structure of proteins 12

R

radioactive gene probes 50
reactions 20, 33
receptor molecules 4
receptor proteins 6
recombinant DNA 44, 45, 47
red blood cells 22, 25, 49
replication of DNA 39
reproduction 56, 57
reproductive cells 36
resolution 3
respiration 31
respiratory gases 22
restriction enzymes 44
Rf values 17
ribonucleic acid 38
ribose sugar 38
ribosomes 4, 42
RNA 4, 38, 42
RNA polymerase 42
rough endoplasmic reticulum 4

S

saturated lipids 14
scanning electron microscope (SEM) 3
secondary response 49
secondary structure of proteins 12
selective breeding 57
semi-conservative replication 39
semilunar valves 28
sino-atrial node (SAN) 28
smooth endoplasmic reticulum 4
smooth muscle 26
specialised cells 23
starch 11, 16
start codons 40
stomata 52
stop codons 40
stroke volume 30
stroma 5
sugar molecules 10
sugars 11
surface area 23
synchronising breeding 57
systems 22

T

telophase 34
tertiary structure of proteins 12
testosterone 57
thermostable enzymes 32, 33
thylakoid membranes 5
thymine 38
tidal volume 30
tissue fluid 25
tissues 22, 24
toxins 30, 48
transcription 42
transformed bacteria 46
transmission electron microscope (TEM) 3
transport across membranes 8, 9
triglycerides 14
tRNA 41, 42
turgid cells 8

U

ultracentrifugation 5
unsaturated lipids 14

V

vagus nerve impulses 30
valves 24, 28, 29
vectors 44, 45
veins 24
vena cava 29
ventilation 26, 27
ventilation cycle 26
ventricles 28, 29
vesicles 4, 9
volume ratio 23

W

water potential 7
waxes 14
white blood cells 22

Z

zygote 36